Atlas of Design

Vol. 4 »

editors
Alethea Steingisser
Caroline Rose
Lauren Tierney

published by
nacis
the North American
Cartographic Information Society

Atlas of Design, Volume Four

Second Printing, July, 2023

Editors: Alethea Steingisser, Caroline Rose, and Lauren Tierney

Published by the North American Cartographic Information Society, *nacis.org*

A qui appartient l'Arctique? Des revendications multiples. (Who owns the Arctic? Multiple claims.) is taken from *L 'atlas des frontières* by Delphine Papin, Bruno Tertrais, and Xemartin Laborde; les Arènes, Paris, 2016.

The type in this book has been set in Avenir, by Adrian Frutiger.

Photos of the *Atlas of Design* are courtesy of Caroline Rose.

Printed in the United States of America by John Roberts, Inc., Minneapolis, MN.

ISBN 979-8-9866856-4-9

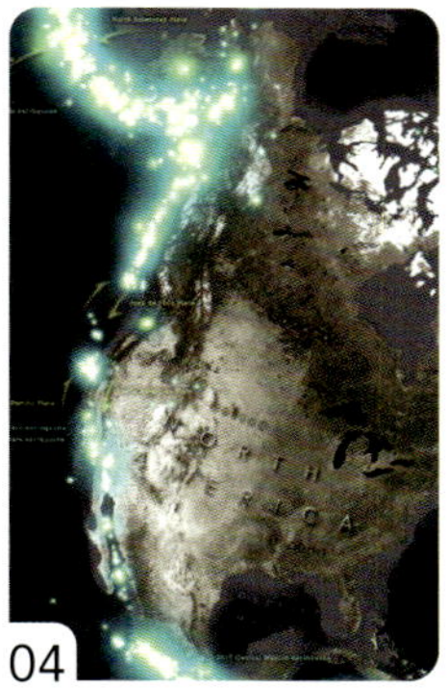
04

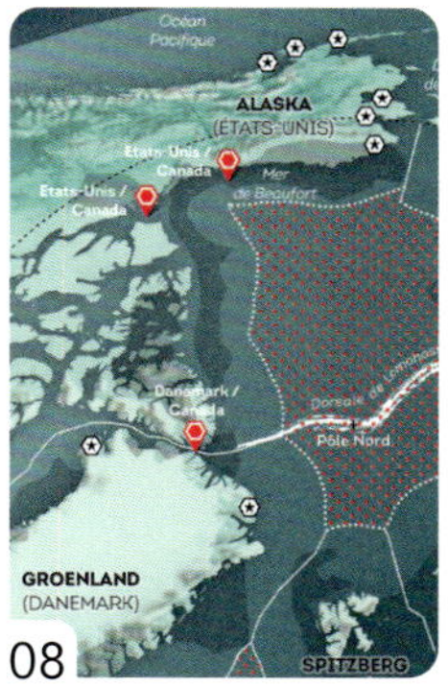

08

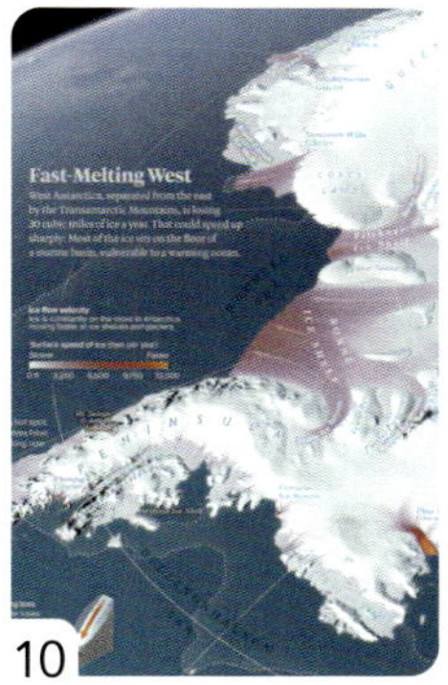

10

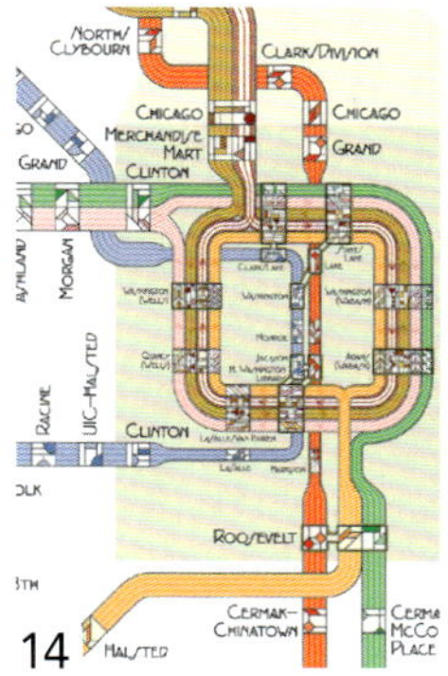

14

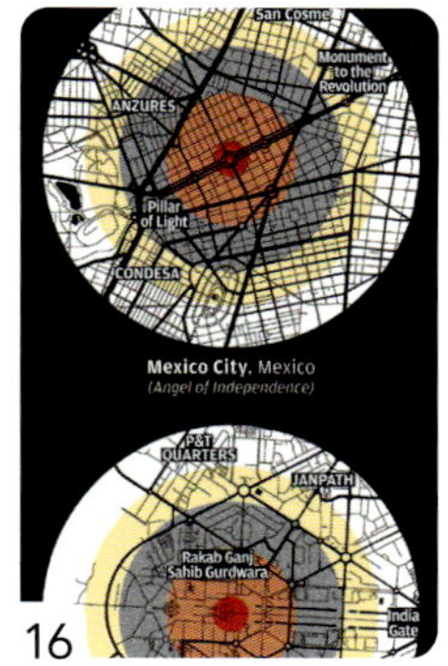

16

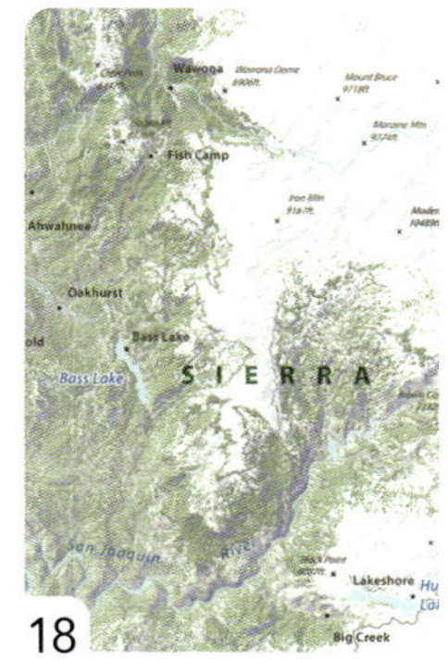

18

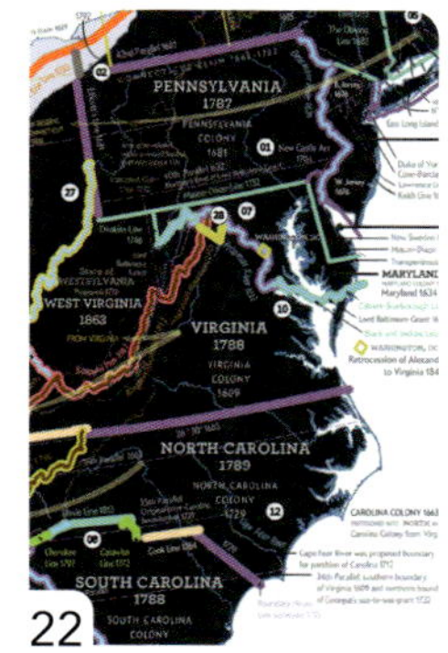

22

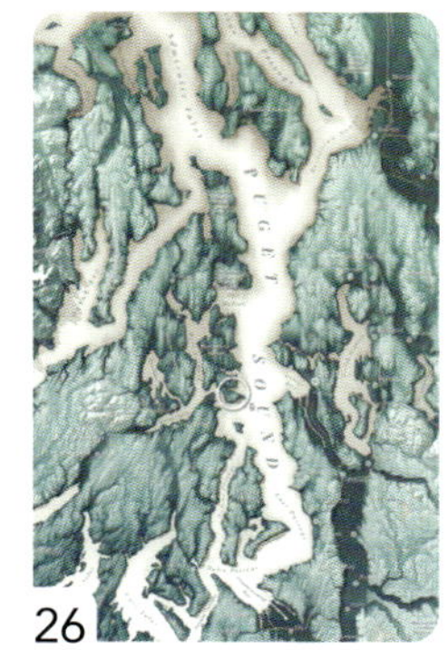
26

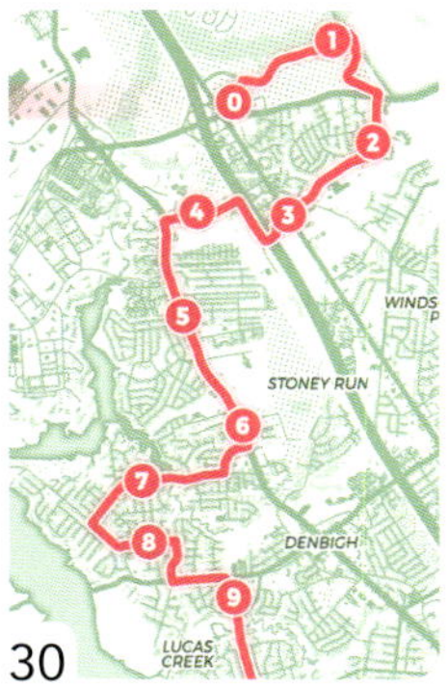

30

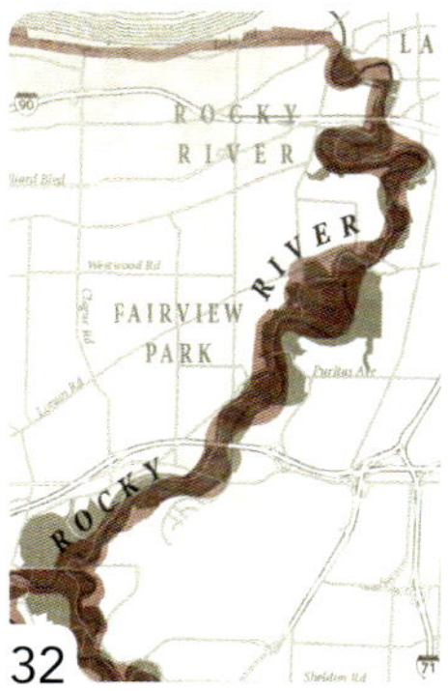

32

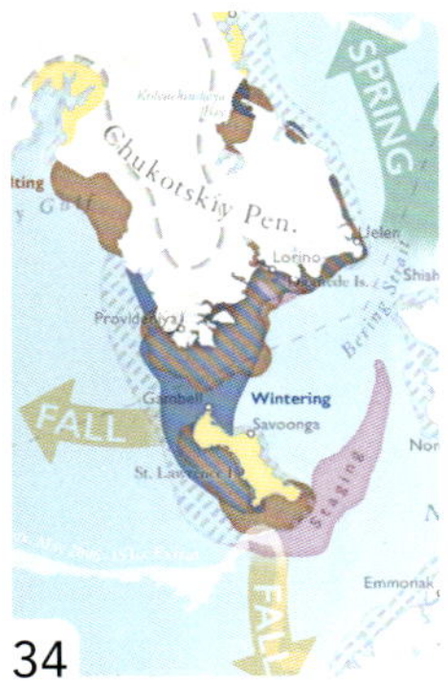

34

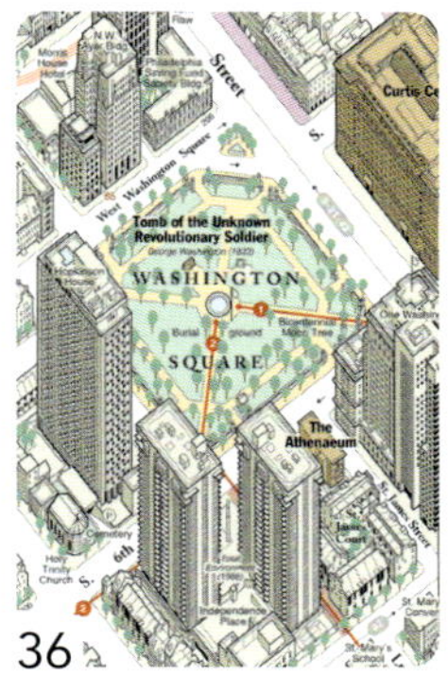

36

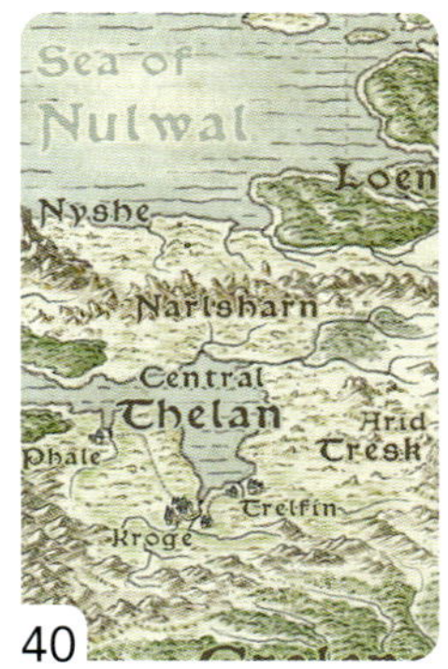

40

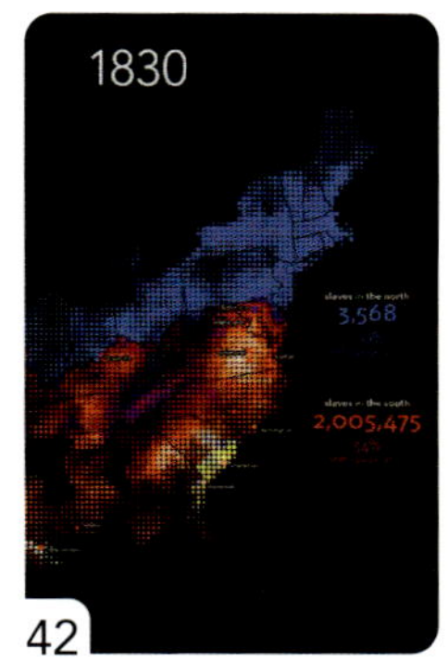

42

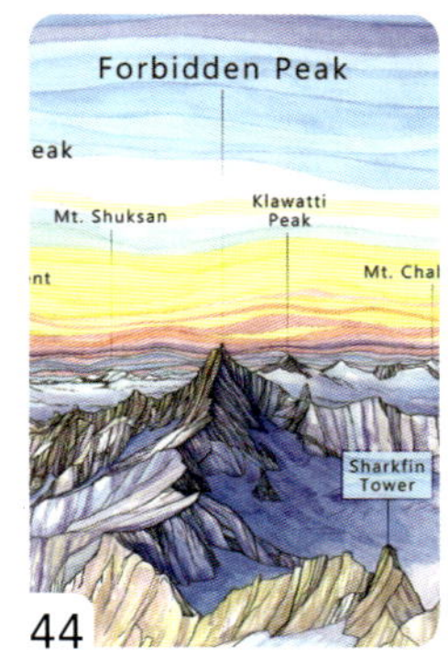

44

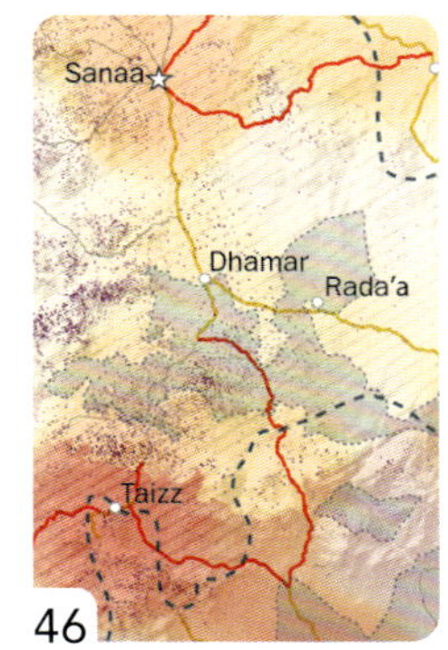

46

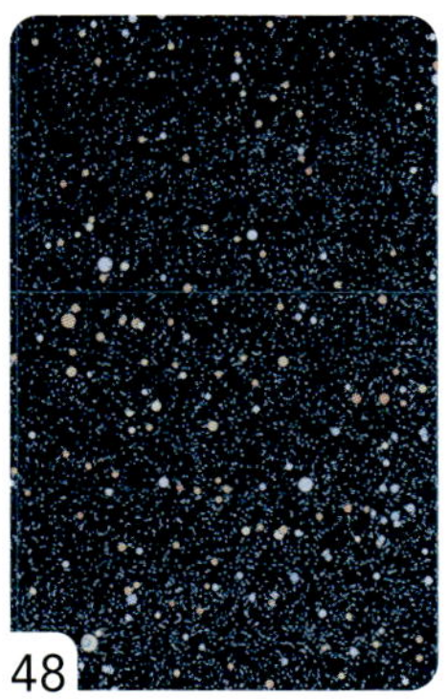
48

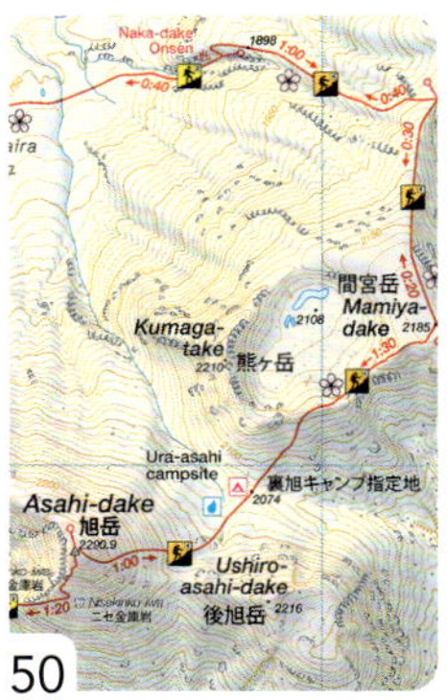

50

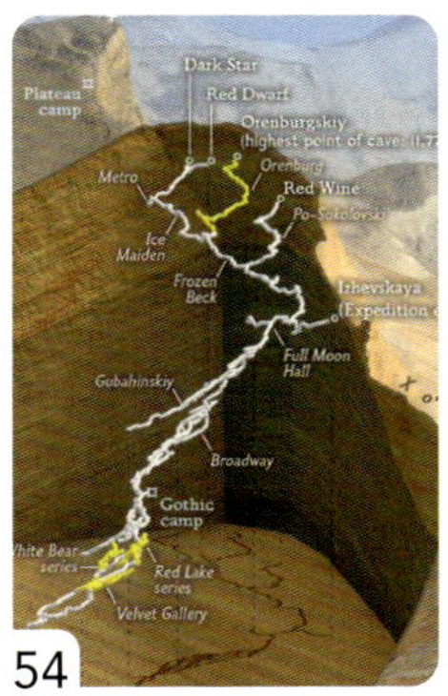
54

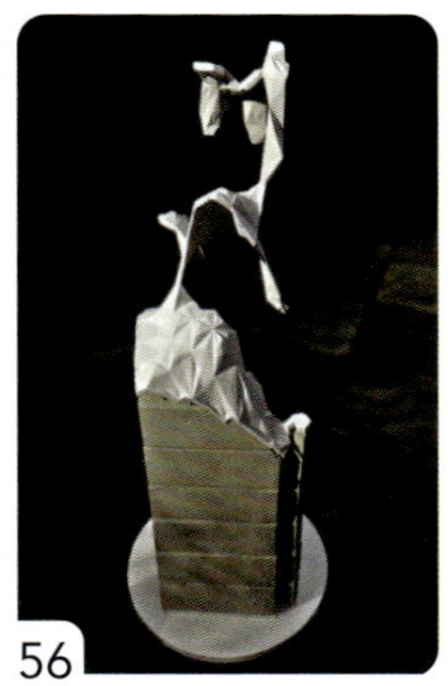
56

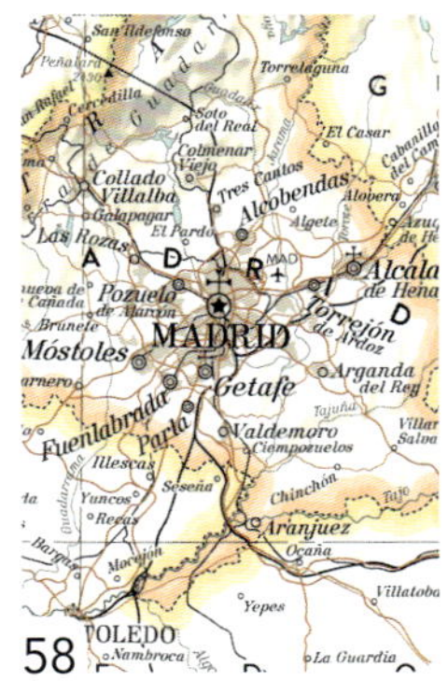

58

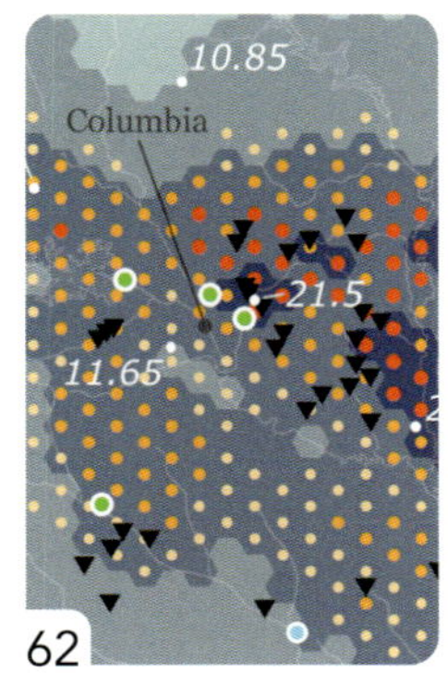

62

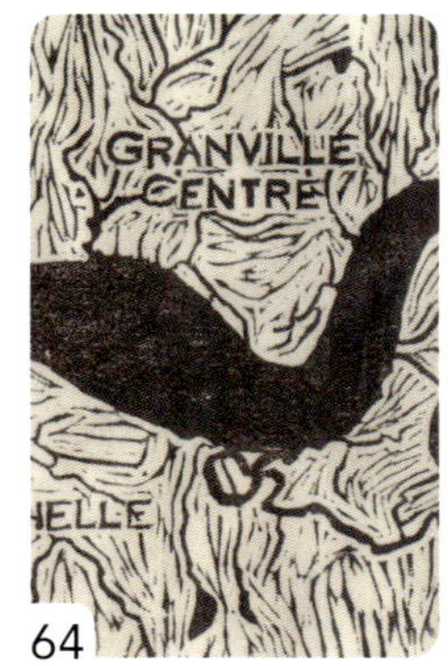

64

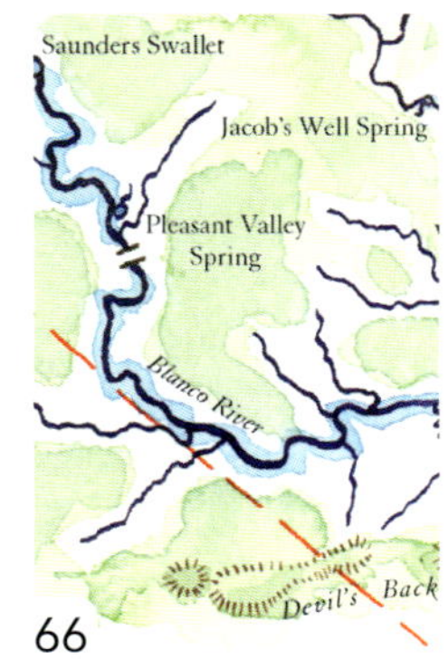

66

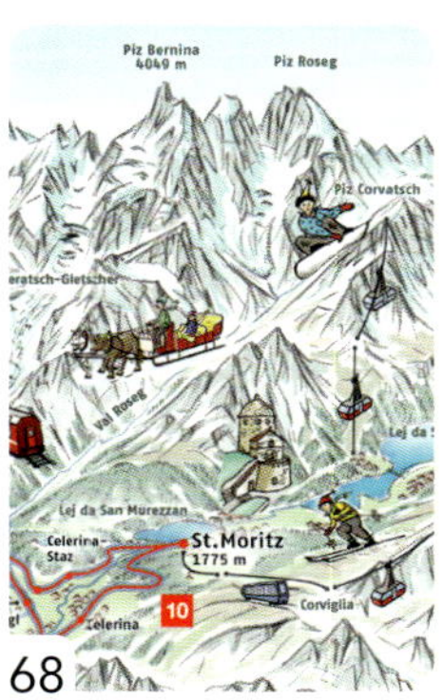

68

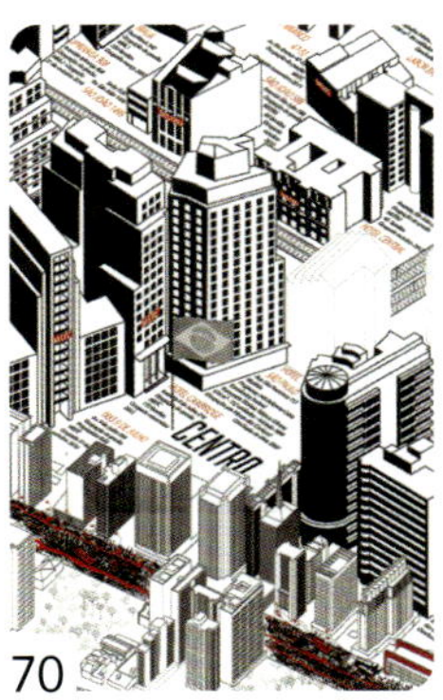
70

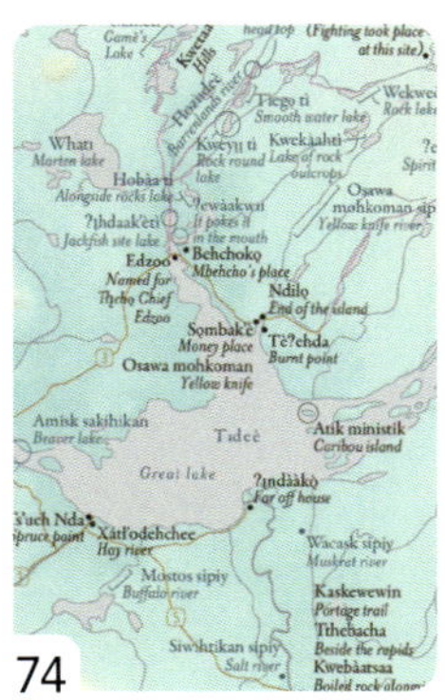
74

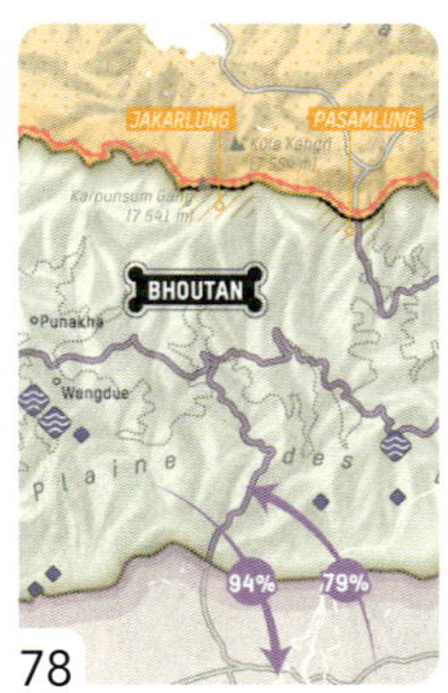

78

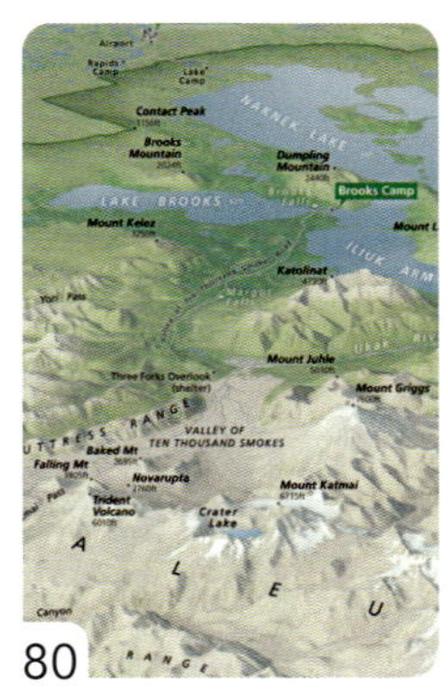
80

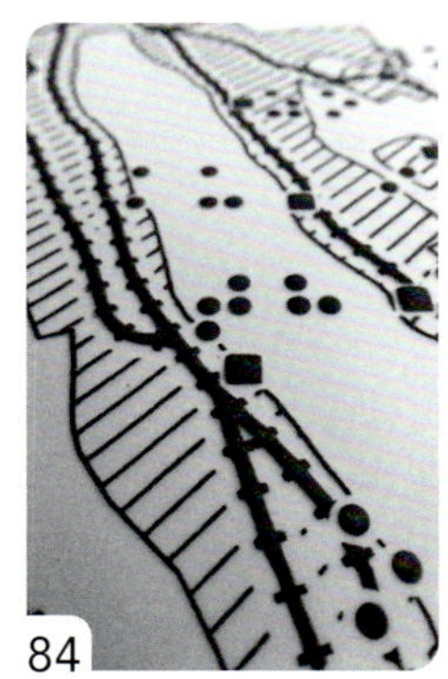
84

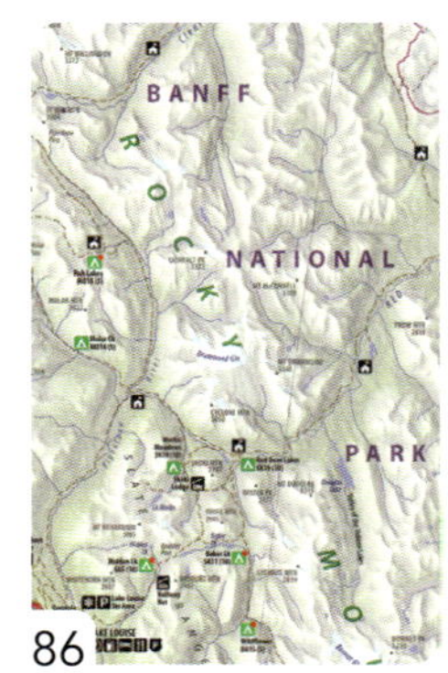

86

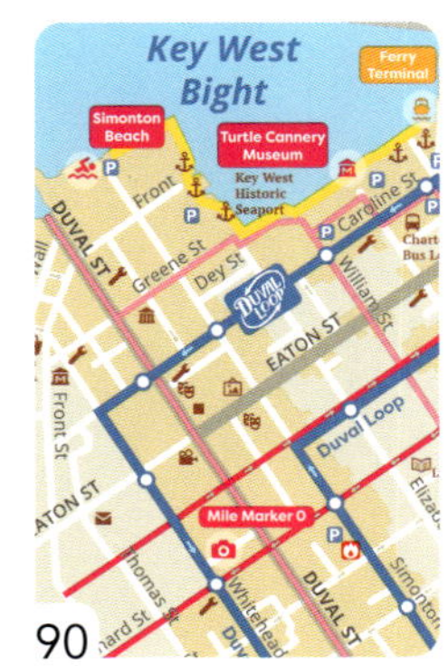

90

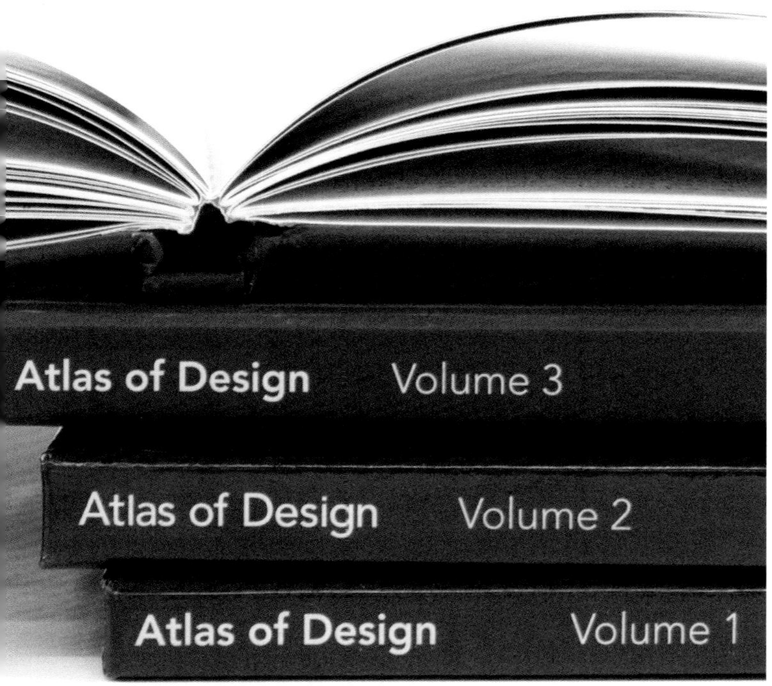

nacis

The *Atlas of Design* is a NACIS publication. The North American Cartographic Information Society improves communication, coordination, and cooperation among the producers, disseminators, curators, and users of cartographic information. The organization is founded on education and information sharing, as well as promoting cartographers' works through publications such as the *Atlas*.

NACIS hosts an annual conference for professionals and enthusiasts from government, commercial, and not-for-profit groups along with academics, scholars, map/GIS librarians, artists, technologists, and students interested in the world of maps and geographic information. To learn more about NACIS visit nacis.org.

A Note from the Editors

We are proud to continue the *Atlas of Design* series with a fourth volume of beautiful and compelling maps.

As professional cartographers ourselves, we reference the previous volumes for guidance when struggling with layout, labels, colors, and other cartographic challenges. We pore over the pages to gain a deeper understanding of our world. We gift copies to friends and relatives to help them understand what cartographers do, and to illustrate the diversity of our field and the power of a skillfully crafted map.

As editors for this volume, we are overwhelmed by the breadth of our profession. We are inspired by the vision, talent, and passion of our fellow cartographers. And we are honored to showcase contemporary cartography from around the globe.

This book is a collaborative effort that has relied on the labor of volunteers and the collective expertise of an extensive network of people we'd like to acknowledge:

The board of the North American Cartographic Information Society (NACIS) provides resources to produce the *Atlas* every two years. We are strongly connected to the NACIS community and are thankful for the opportunity to serve the organization as editors.

We are grateful to the previous editorial teams who established a visionary and practical foundation for this project. We thank the previous editors—Ginny Mason, Sam Matthews, Marty Elmer, and Tim Wallace—for their support as they passed the torch to us. We are especially indebted to former editor Daniel Huffman for his patience and guidance, as well as his unflagging passion for sharing cartographic wisdom.

We could not create the *Atlas* without the maps submitted by cartographers from around the globe. We thank every cartographer, whether selected for inclusion in this volume or not, for sharing their work and giving us a wealth of powerful and beautiful maps from which to choose. Over 300 maps were submitted, each the result of dozens or even hundreds of hours of research and design. Narrowing down that mountain of maps was no easy feat. We were guided by a panel of judges who donated many hours to evaluate the submissions. We extend a heartfelt thank you to the judges for this volume: Tanya Buckingham, Juliane Cron, Daniel Huffman, Lauren James, Hans van der Maarel, Joanna Merson, John Nelson, Delphine Papin, Tom Patterson, Robert Roth, Alex Tait, and Tim Wallace.

We are not experts in writing, so we relied on the expertise of our copy editor, Shelley Sperry, who helped bring structure and consistency to the text with her attention to detail and talent for wordsmithing.

The InfoGraphics Lab in the Department of Geography at the University of Oregon supported the production of this volume by providing time and resources to create the book layout.

The American Printing Company has been our partner since the first volume, and we continue to rely on their expertise to make our vision a reality.

Last but not at all least, we are grateful to you, our readers, for your curiosity and interest in cartography, and for listening to the stories these maps have to tell. Thank you for supporting the *Atlas of Design*.

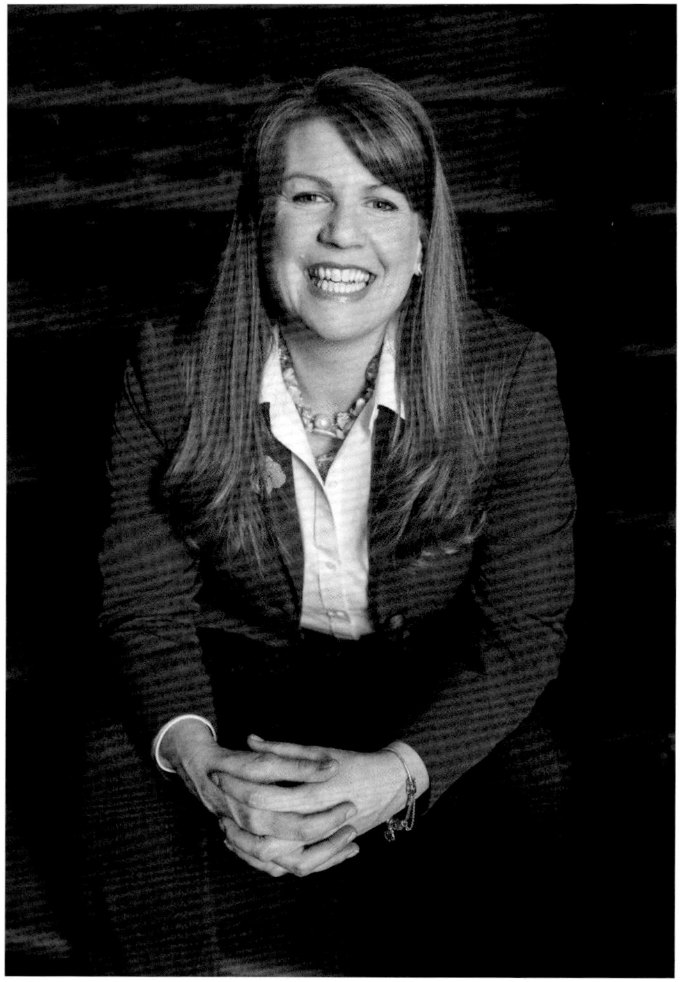

Introduction by Tanya Buckingham

Tanya is the creative director at the University of Wisconsin Cartography Lab, where she trains students in the middle space between academia and industry. She has served as president and executive director of NACIS and was production cartographer at XNR Productions. Tanya is most grateful for all of the people that maps have brought into her life during her eighteen-year career—especially the students she gets to work with every day.

A Noble Craft

As our fingers trail along the Himalaya on a map,
we may not feel the thin air in our lungs. We may not feel the
ground beneath us tremble as the ice cleaves, causing our maps to shift with
the thaw. And we may not have felt the truck lurch over destroyed roads in a
country rocked by violence.

Yet, we can know them to be true
for places we have been,
and imagine them for unknown places.
And we can understand so much about what we can show
and what we cannot.

We can represent unequal distribution of resources
wealth there
internet here
air there
water here
peace there
protests here
famine there
pollution here
people here.

How we move those resources or contaminants
from one place to another
Or, how we shape our experiences
within the grooves upon the canvas which we were given

A reminder that we all place our feet upon the same ground
That my actions impact you, and yours me
That what we cannot see is equally important to what we can

A map is a tool that helps us make connections
We're careful, though, not to assume that one is necessari y
the result of the other
but to inquire,
what is the result of these intersections

And, that what we cannot see is as powerful as what we can;
that their interaction may be beyond our comprehension

Some would argue the map is more than it really is
But, can we not celebrate it for all that it is?
simply
Without trying to force it to be more than it should
a tool
With a purpose of what?
empathy.

To understand, and to feel that different circumstances
have shaped your life, and mine
And, these differences have
exploited
enforced
killed for
shared
created peace

It can be
the hammer that drives the final nail in the fundraising pitch
the artifact that for the first time elucidates
the phenomenon a researcher has studied for a decade
the lens for a child to understand that she is part of something
so much bigger than herself

And, we are careful
not to assume that one is necessarily the result of the other
to know that the sum is not the individual
And, that what we cannot see is as powerful as what we can;
that their interaction may be beyond our comprehension

Those who paint these pictures,
situated in space,
bear the weight of great responsibility
to design
to inspire
to capture imagination and attention
to persuade
to be accurate
to be neutral
or not.

If we are still
and we listen

We can hear the cartographer and what is represented
through the medium of the map

We are all connected,
and your action impacts me,
and mine you

That my feet touch the same planet as yours,
and—for now—it's the only home we have

This is why the map is believed to be bigger than it is.
There was a time I thought this was vanity
now, I understand it as hope.

A hope that through our map-making we can
intersect just the right data,
with the most thought-provoking design and capture
the policy-maker's attention
the child's potential
the lost traveler's way

Hope, that we see ourselves as part of a global system,
and that we inspire others to do the same
using the tool that we know can spread empathy.
Empathy. Something we cannot show on a map
in data
and numbers,
in this shade,
or that hue

Whether you call our craft an art or a science,
or our product a data visualization,
propaganda,
a proposal,
a tool,
or something else entirely.

A commitment to empathy, through the vessel of our industry,
is to participate in a noble effort.

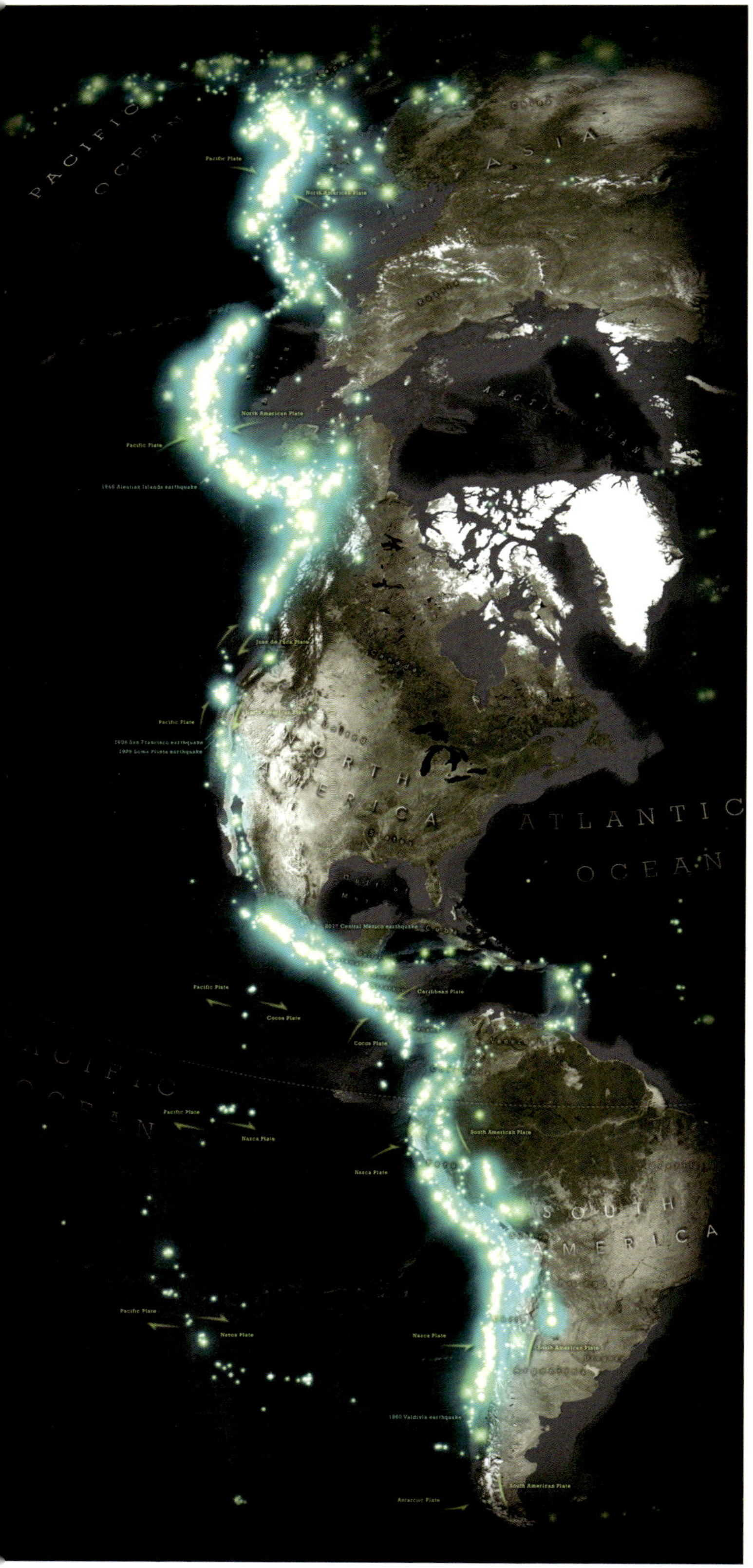

> *Making maps is a great adventure. Each map says something about the phenomenon, but also says something about its maker.*

« Eastern Pacific Ring of Fire

John Nelson

This map was created as part of a larger story about *emergence*. Emergence is the phenomenon of massive collections of independent events—such as individual earthquakes—revealing a larger, previously unknown structure when seen in aggregate. In this map, the larger structure is the system of fractious boundaries between Earth's tectonic plates, and it is beautiful.

Years ago I made another map of historic earthquakes and showed it to my ten-year-old son, Bear. He looked at it for a moment and then said matter-of-factly, "Oh, it's the Ring of Fire." I was surprised by his awareness of earth science at such a young age and equally surprised that I had not seen the pattern myself. I only saw a map of earthquakes. Bear's observation had a huge influence on my work going forward.

These quakes trace a long circuitous chain of fault lines. They, and the global movements they evidence, are pure energy. The symbology on the map is intended to reflect that vibrancy, glowing bright at epicenters and decaying in intensity at a distance. The boundaries where the quakes coalesce appear as a thin crust splitting as a result of light and heat.

To support the aesthetic of punctuated light, I gave the labels an ethereal illumination that casts faint shadows, longer at the edges. This reinforces the arced surface of the Earth, drawing readers into a map with dimension and scale. This was my favorite part of the process because I typically shy away from labeling in my work.

Making maps is a great adventure. I love and appreciate the process more with each one I'm honored to create. Each map says something about the phenomenon, but also says something about its maker. Each map is a unique creation, but also a step in a larger journey. Looking back at the maps I have created over the years, I feel great satisfaction in the evidence of my own evolution in a craft I'm so fortunate to have found.

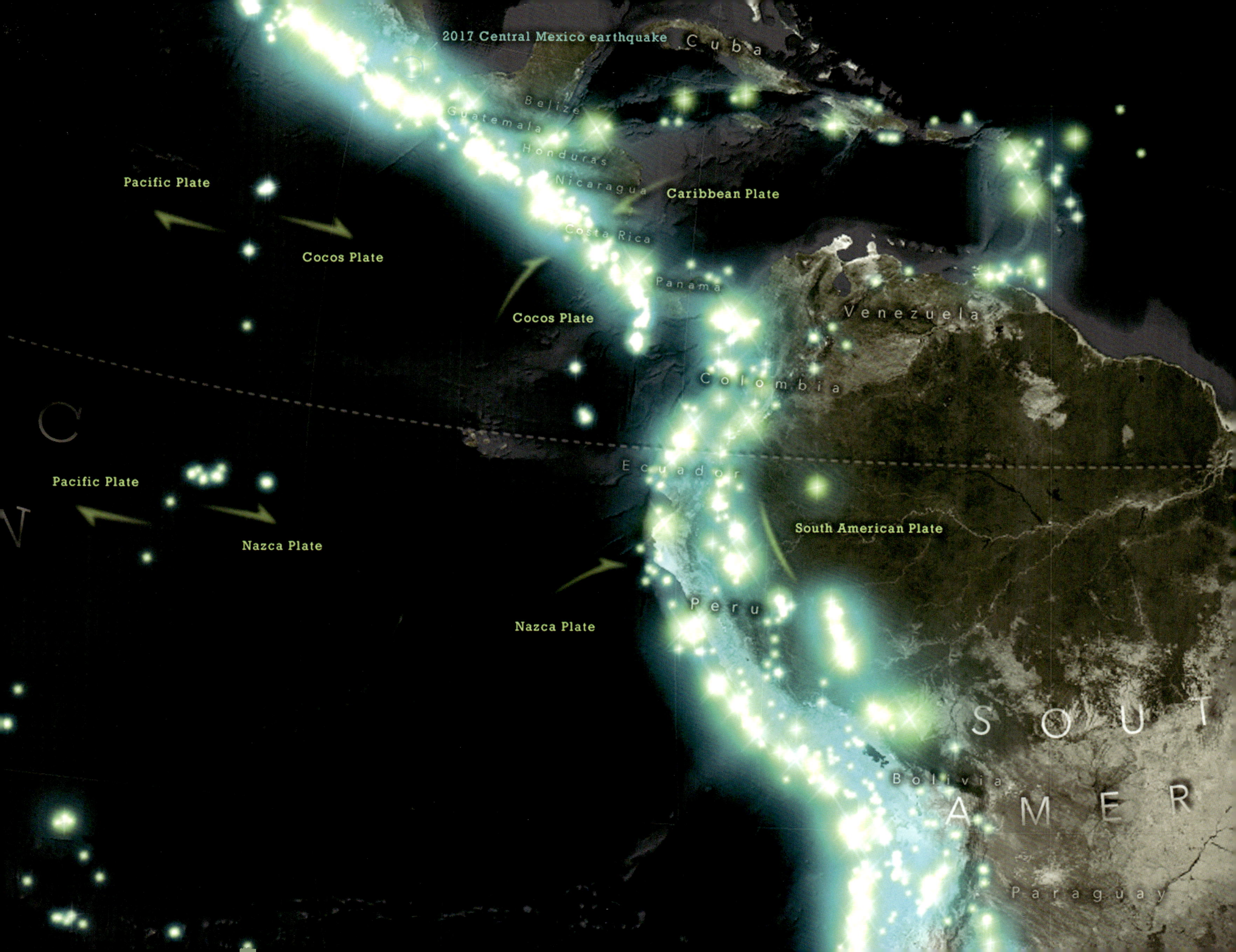
2017 Central Mexico earthquake
Cuba
Belize
Guatemala
Honduras
Nicaragua
Caribbean Plate
Pacific Plate
Cocos Plate
Costa Rica
Panama
Cocos Plate
Venezuela
Colombia
Ecuador
Pacific Plate
Nazca Plate
South American Plate
Peru
Nazca Plate
Bolivia
SOUTH AMER
Paraguay

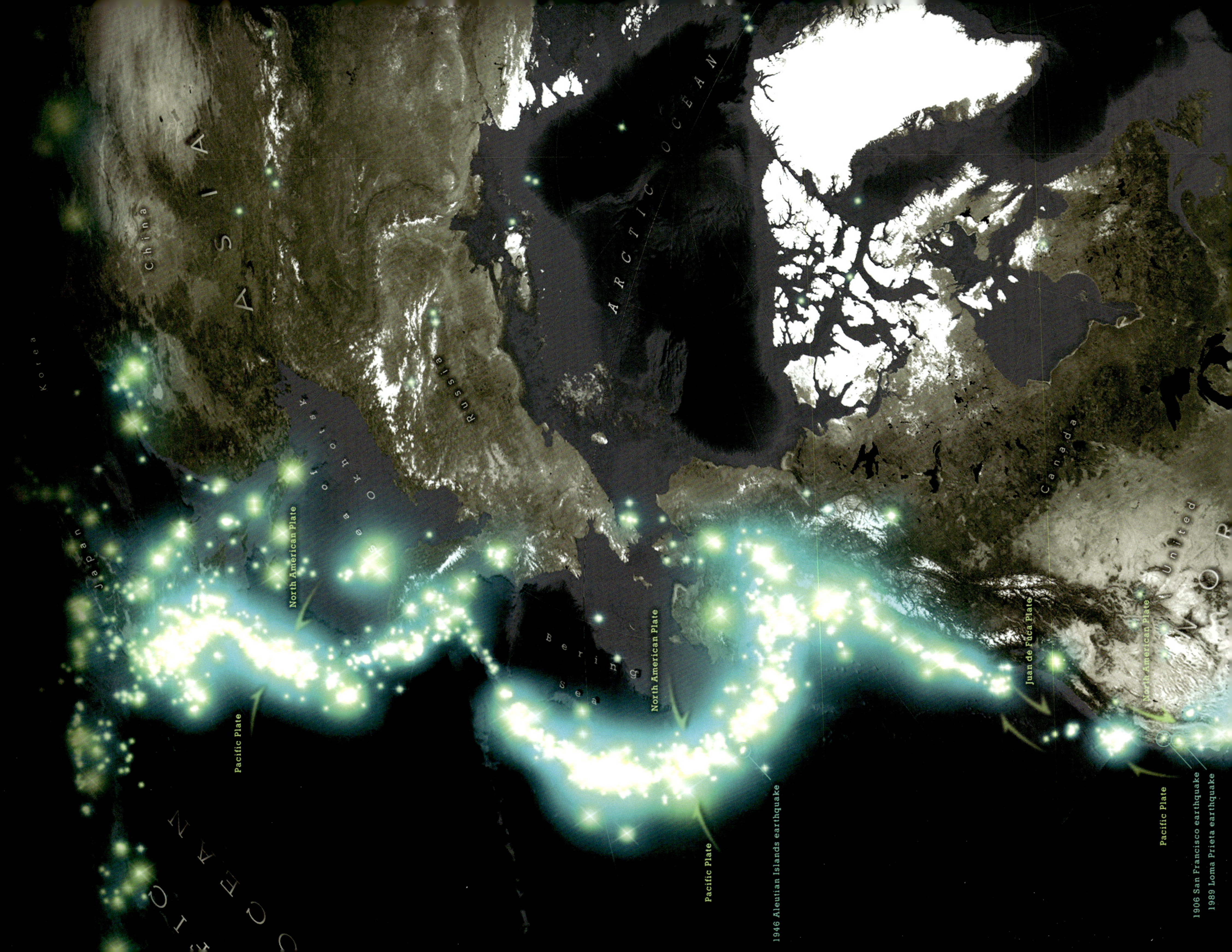
ASIA
China
Korea
Russia
Sea of Okhotsk
ARCTIC OCEAN
Canada
United
Japan
North American Plate
Pacific Plate
Bering Sea
North American Plate
Pacific Plate
1946 Aleutian Islands earthquake
Juan de Fuca Plate
North American Plate
Pacific Plate
1906 San Francisco earthquake
1989 Loma Prieta earthquake

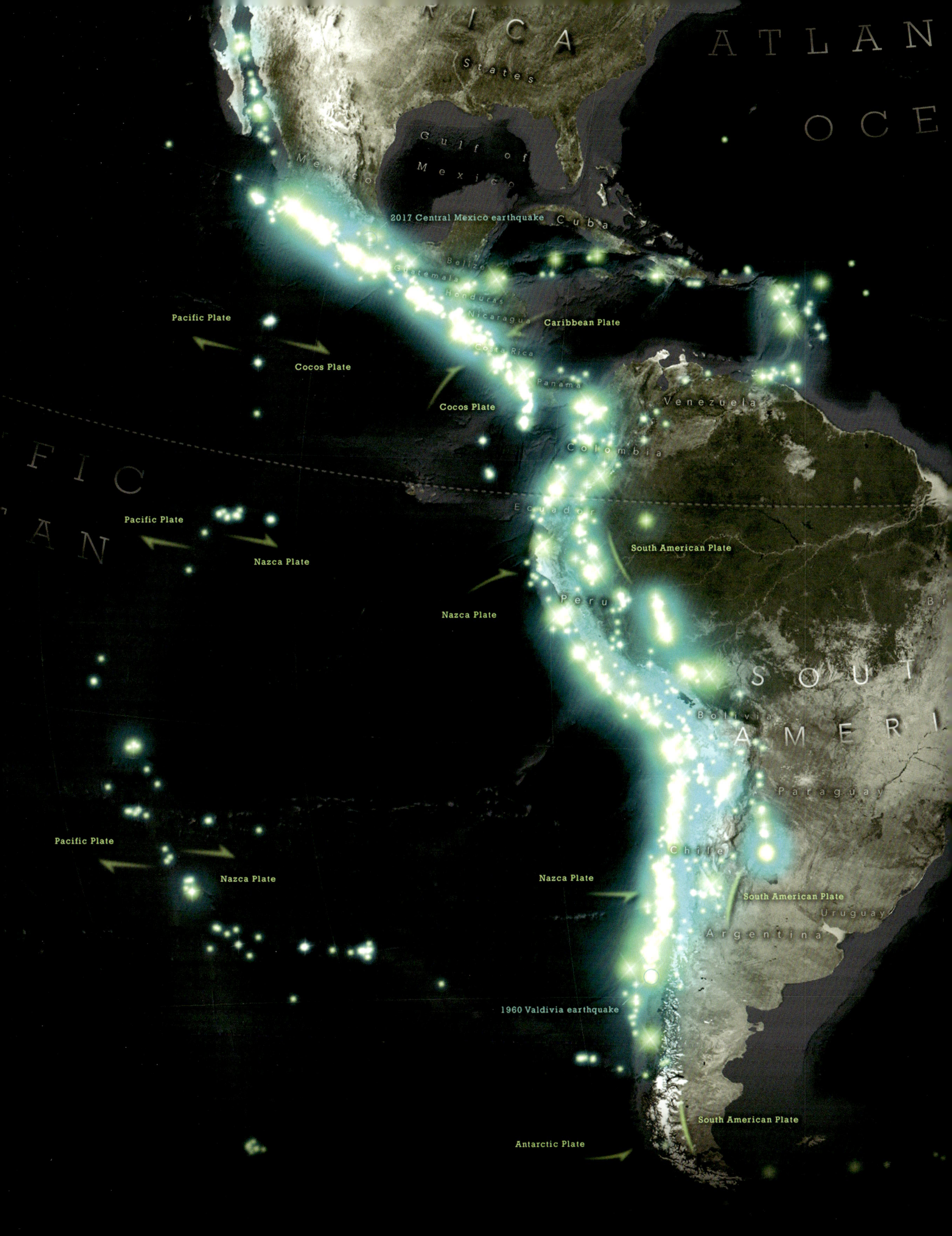
States
Gulf of Mexico
Mexico
2017 Central Mexico earthquake
Cuba
Belize
Guatemala
Honduras
Nicaragua
Caribbean Plate
Costa Rica
Panama
Pacific Plate
Cocos Plate
Cocos Plate
Venezuela
Colombia
Ecuador
Pacific Plate
Nazca Plate
South American Plate
Peru
Nazca Plate
Bolivia
Paraguay
Chile
Pacific Plate
Nazca Plate
Nazca Plate
South American Plate
Uruguay
Argentina
1960 Valdivia earthquake
South American Plate
Antarctic Plate

9 À QUI APPARTIENT L'ARCTIQUE ?

DES REVENDICATIONS MULTIPLES

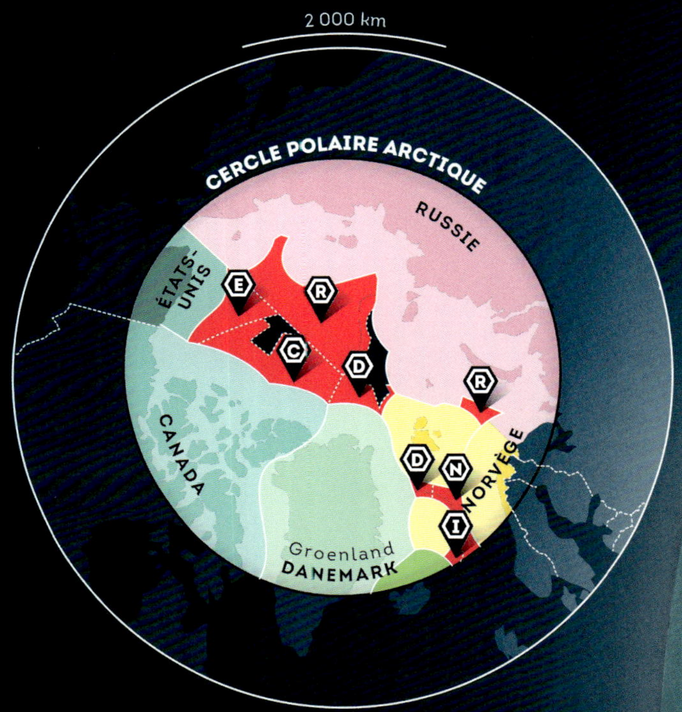

TERRITOIRES ET REVENDICATIONS

- Frontière reconnue
- Limites des ZEE (200 milles marins)

	Rus.	Nor.	Isl.	Dan.	Can.	É-U
Zone économique exclusive						
Plateau continental revendiqué	R	N	I	D	C	E

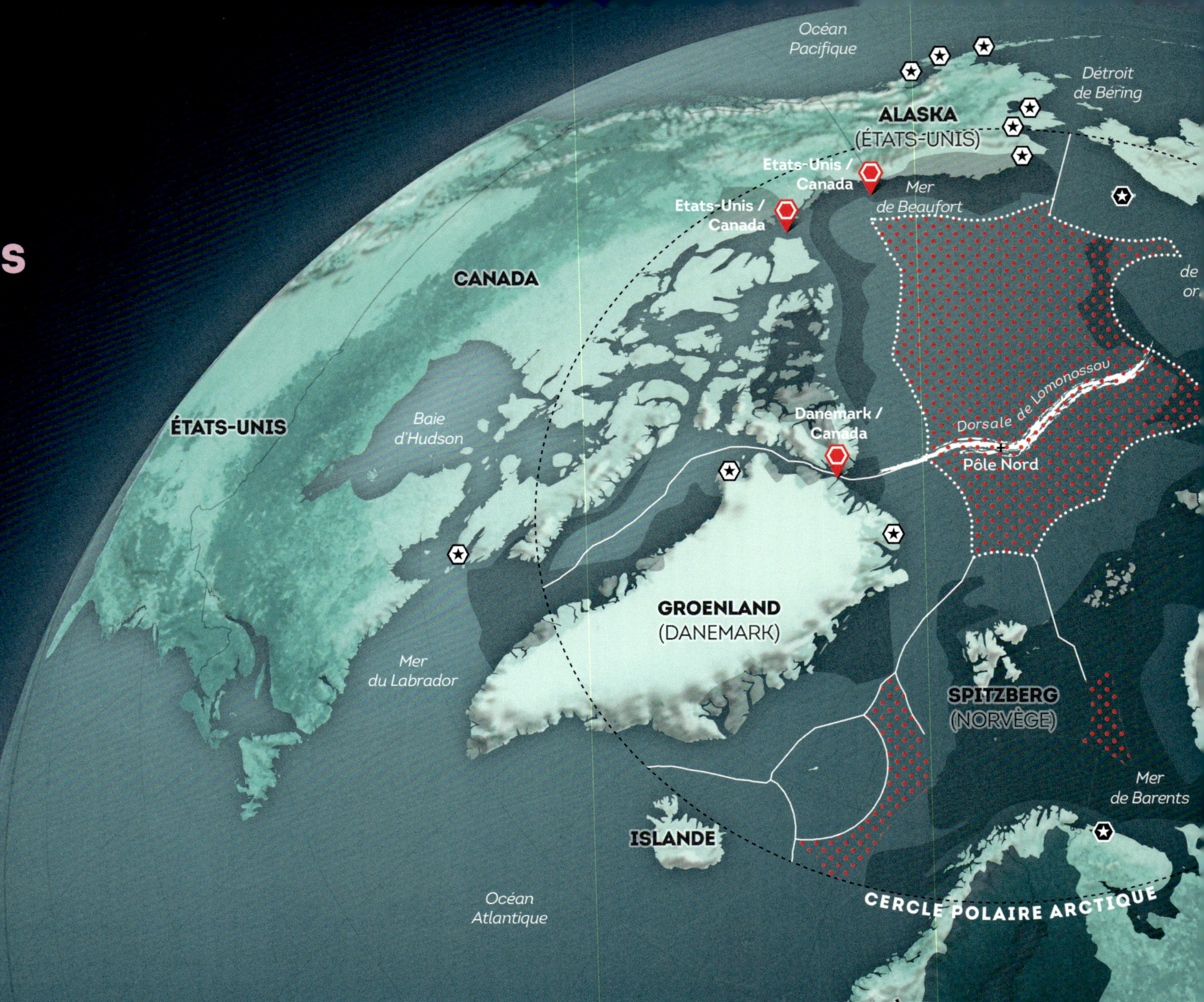

PRÉSENCE MILITAIRE ET REVENDICATIONS

- Frontière reconnue
- Haute mer (zone internationale)
- Zone de revendication
- Base militaire russe
- Base militaire américaine
- Différends territoriaux
- Présence avérée ou potentielle d'hydrocarbures

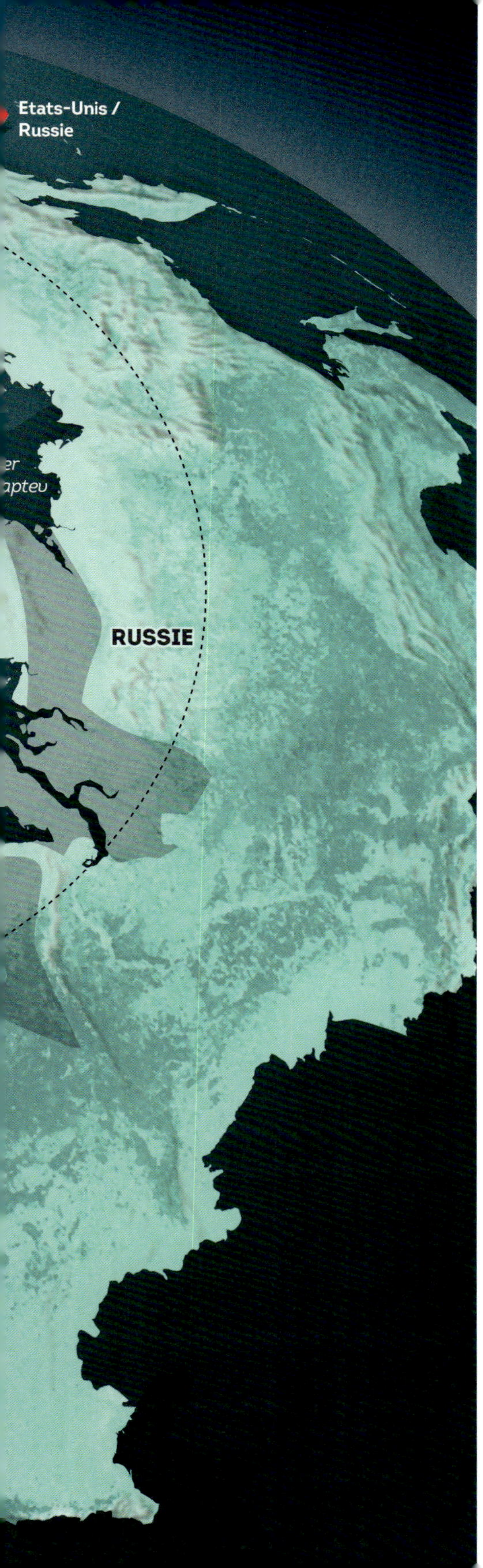

« *À Qui Appartient l'Arctique? Des Revendications Multiples,* Who Owns the Arctic? Multiple Claims

Xemartin Laborde and Delphine Papin — Le Monde, with Bruno Tertrais

This map is from the *Atlas of Borders*, a cartographic and geopolitical guide published in France in 2016. The book describes a variety of contemporary challenges along international boundaries, including colonial borders, invisible borders, burning borders, and walled borders. In "Who Owns the Arctic?" we invite readers to the North Pole, where many nations lay claim to the same territories.

The primary design choice for this panoramic map is a duotone effect. In this cold, nocturnal landscape the color blue dominates—azure and clear for glaciers and continents, gloomier and darker for the sky and the oceans. The blue contrasts sharply with the red that marks geopolitical tensions, territorial disputes, and the areas claimed in these disputes.

The map is an orthographic projection, displaying the planet in its true spherical form as viewed from space. A halo on the horizon provides a sense of realism. Imagery depicting forested areas is draped over the continents, again in shades of blue, adding texture. The varied bluish tints identify and highlight the countries bordering the Arctic Ocean, with other countries remaining in the dark.

The color red marks disputed areas and recognized boundaries are represented in white, a color chosen because it is associated with the neutrality of international agreements. The graticule and the Arctic Circle are gray and black. A locator map helps viewers understand the complex diplomatic context, distinguishing countries in the polar projection with a rainbow of colors and small pins arranged to show disputed territories. All of the elements in the legend align with the graphic principles within the rest of the atlas.

Our map emphasizes the tensions and competing land claims in a region that only appears to be empty and peaceful from afar.

THE MELTING OF ANTARCTICA

Snow falling on Antarctica through the ages has piled up in domes of ice more than two miles thick. Most of that is likely safe for many centuries. But as the climate warms, the edges of the frozen continent are crumbling. The natural flow of ice, down through coastal glaciers and floating ice shelves to the sea, is speeding up. By 2100 the ice loss may contribute several feet to global sea-level rise.

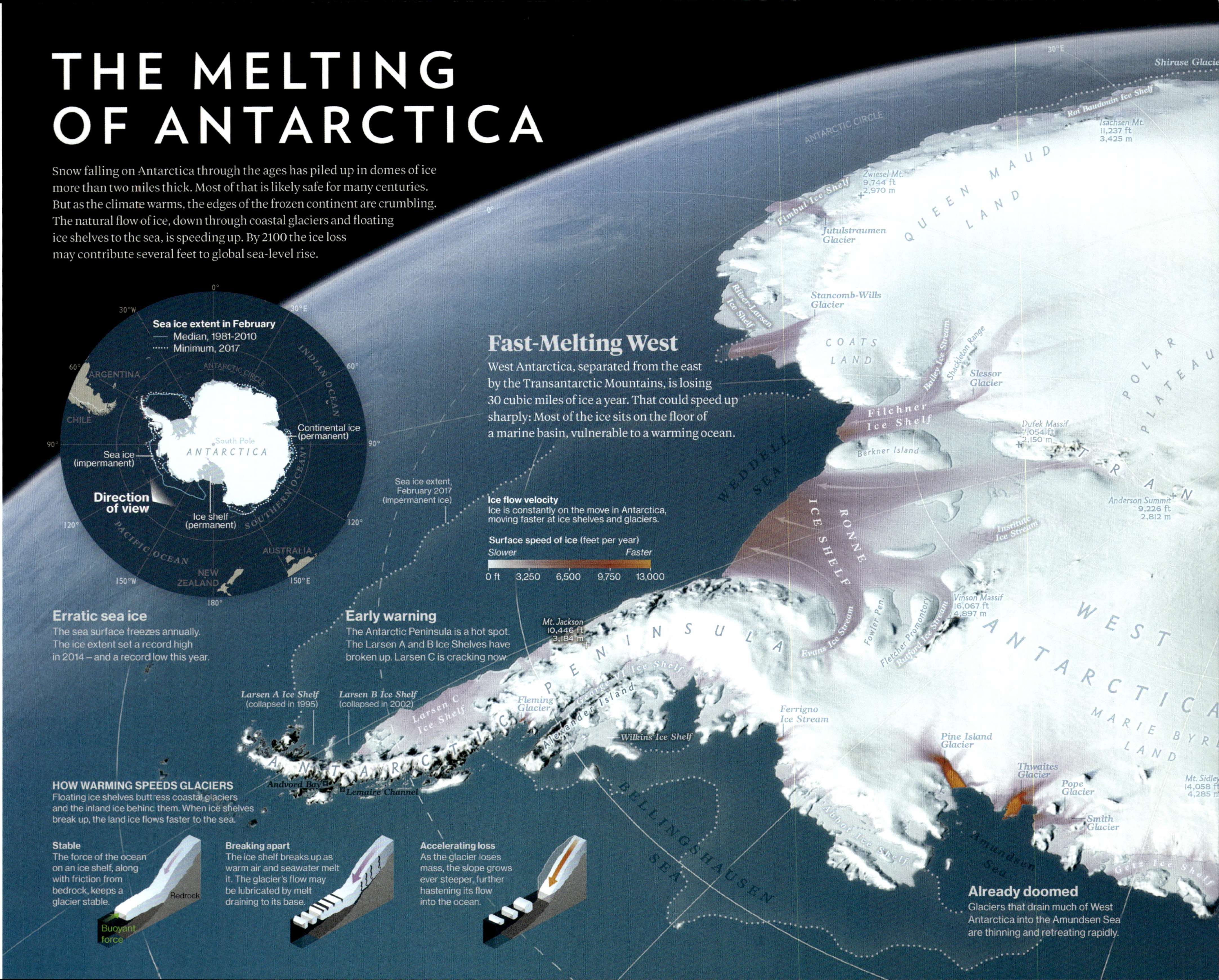

Fast-Melting West

West Antarctica, separated from the east by the Transantarctic Mountains, is losing 30 cubic miles of ice a year. That could speed up sharply: Most of the ice sits on the floor of a marine basin, vulnerable to a warming ocean.

Erratic sea ice

The sea surface freezes annually. The ice extent set a record high in 2014 – and a record low this year.

Early warning

The Antarctic Peninsula is a hot spot. The Larsen A and B Ice Shelves have broken up. Larsen C is cracking now.

HOW WARMING SPEEDS GLACIERS

Floating ice shelves buttress coastal glaciers and the inland ice behind them. When ice shelves break up, the land ice flows faster to the sea.

Already doomed

Glaciers that drain much of West Antarctica into the Amundsen Sea are thinning and retreating rapidly.

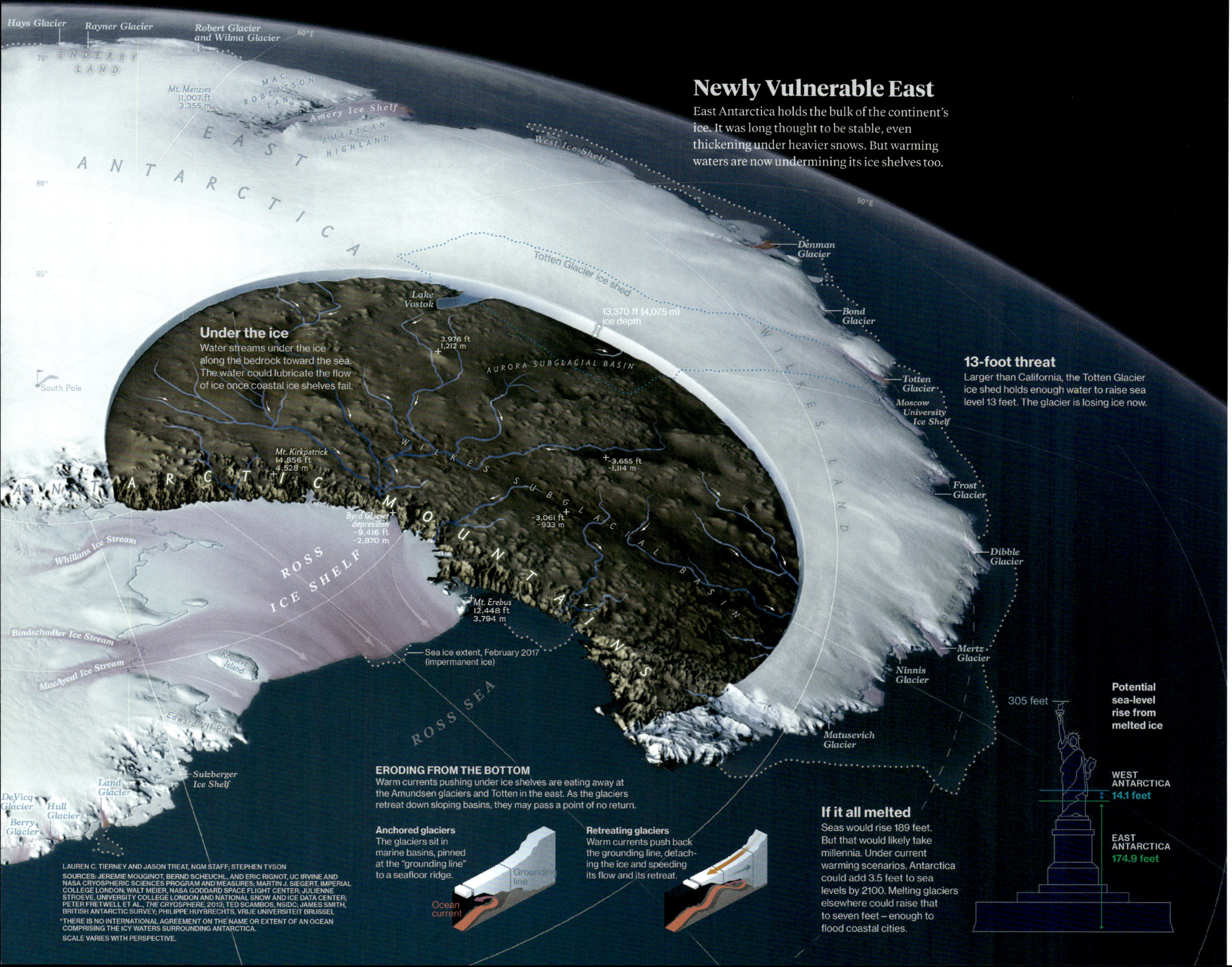

Newly Vulnerable East

East Antarctica holds the bulk of the continent's ice. It was long thought to be stable, even thickening under heavier snows. But warming waters are now undermining its ice shelves too.

Under the ice

Water streams under the ice along the bedrock toward the sea. The water could lubricate the flow of ice once coastal ice shelves fail.

13-foot threat

Larger than California, the Totten Glacier ice shed holds enough water to raise sea level 13 feet. The glacier is losing ice now.

ERODING FROM THE BOTTOM

Warm currents pushing under ice shelves are eating away at the Amundsen glaciers and Totten in the east. As the glaciers retreat down sloping basins, they may pass a point of no return.

Anchored glaciers

The glaciers sit in marine basins, pinned at the "grounding line" to a seafloor ridge.

Retreating glaciers

Warm currents push back the grounding line, detaching the ice and speeding its flow and its retreat.

If it all melted

Seas would rise 189 feet. But that would likely take millennia. Under current warming scenarios, Antarctica could add 3.5 feet to sea levels by 2100. Melting glaciers elsewhere could raise that to seven feet – enough to flood coastal cities.

LAUREN C. TIERNEY AND JASON TREAT, NGM STAFF; STEPHEN TYSON

SOURCES: JEREMIE MOUGINOT, BERND SCHEUCHL, AND ERIC RIGNOT, UC IRVINE AND NASA CRYOSPHERIC SCIENCES PROGRAM AND MEASURES; MARTIN J. SIEGERT, IMPERIAL COLLEGE LONDON; WALT MEIER, NASA GODDARD SPACE FLIGHT CENTER; JULIENNE STROEVE, UNIVERSITY COLLEGE LONDON AND NATIONAL SNOW AND ICE DATA CENTER; PETER FRETWELL ET AL., *THE CRYOSPHERE*, 2013; TED SCAMBOS, NSIDC; JAMES SMITH, BRITISH ANTARCTIC SURVEY; PHILIPPE HUYBRECHTS, VRIJE UNIVERSITEIT BRUSSEL

*THERE IS NO INTERNATIONAL AGREEMENT ON THE NAME OR EXTENT OF AN OCEAN COMPRISING THE ICY WATERS SURROUNDING ANTARCTICA.

SCALE VARIES WITH PERSPECTIVE.

The Melting of Antarctica »

Lauren C. Tierney, Stephen Tyson, and Jason Treat — National Geographic Magazine

For more than 120 years, *National Geographic* cartographers have mapped Antarctica, and the magazine has illustrated the complex processes that occur in this remote landscape. The tradition continued with "The Melting of Antarctica," published in July 2017, a map highlighting the effects of climate change on the continent.

The thematic elements of the map include four key effects of climate change: 1) A comparison of the western and eastern sides of the continent shows that the west is changing more rapidly. 2) An analysis of ice velocity demonstrates that it is constantly on the move and that movement at the edges of the continent has been speeding up. 3) An illustration of subglacial rivers and lakes shows the flow of water through bedrock, which may lubricate ice movement in the future. 4) A look at the dynamics of the ice shelves that hold back glacial ice shows changes happening as a result of warming waters.

The final design of "The Melting of Antarctica" in both print and digital forms required months of extensive fine-tuning and incorporation of cartographic elements from historic *National Geographic* maps, tying our modern version to the magazine's heritage of mapping science and exploration. One of our favorite touches was the "golf flag" depicting the South Pole, which was inspired by an illustrated map of Greenland, published in the January 1956 edition of the magazine. We also simplified the Statue of Liberty graphic showing the potential depth of sea level rise to match the overall feel of the map, blended the map labels with the landscape, adjusted the lighting and shadows to emphasize the volume of the ice, and revised the final purple-red-orange color ramp for the ice flow velocity. We even revised the map's title three times.

From initial concept to final design, our team took six months to plan, research, design, and do all the necessary cartographic tweaking to accurately communicate the dire effects of climate change at the bottom of the world.

Fast-Melting West

Vest Antarctica, separated from the east y the Transantarctic Mountains, is losing 0 cubic miles of ice a year. That could speed up harply: Most of the ice sits on the floor of marine basin, vulnerable to a warming ocean.

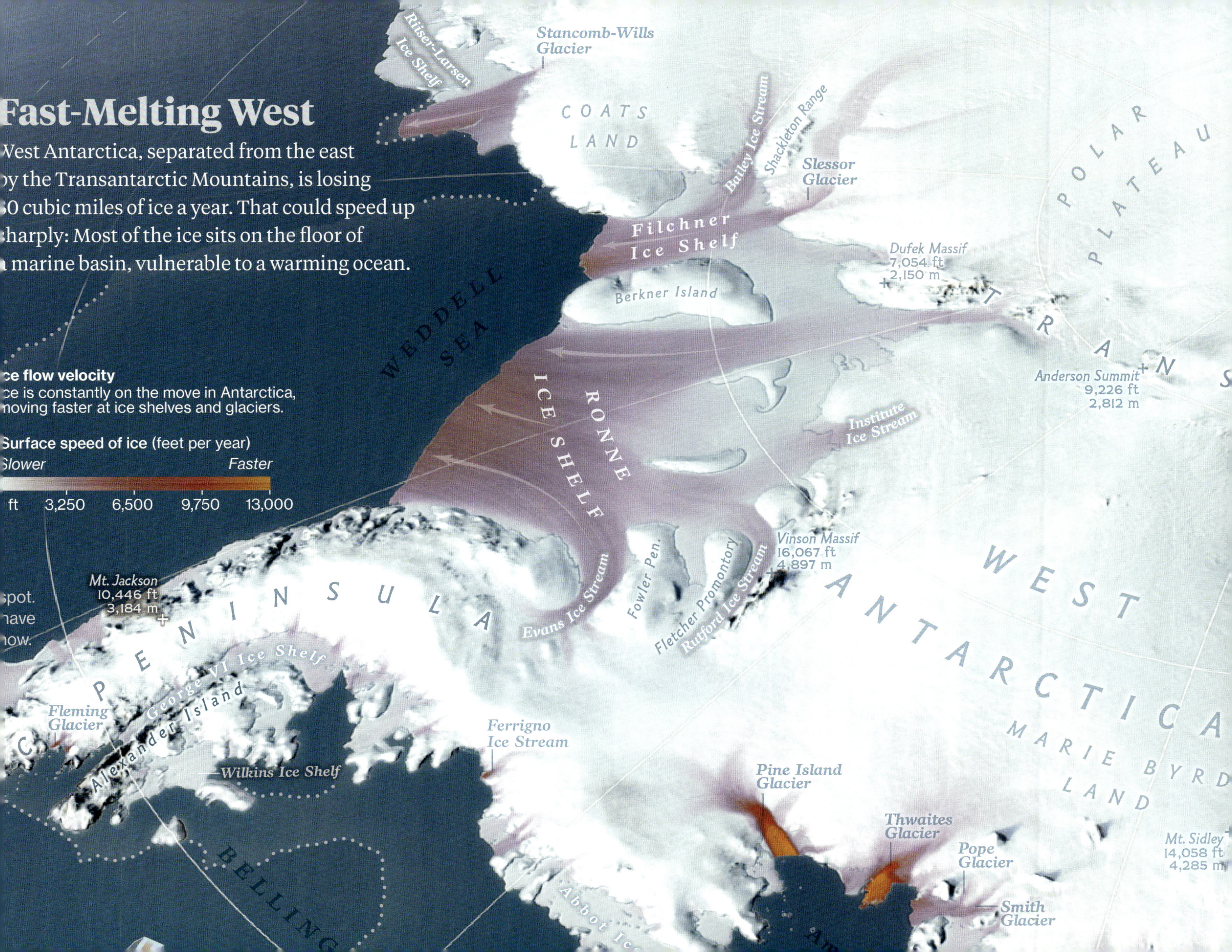

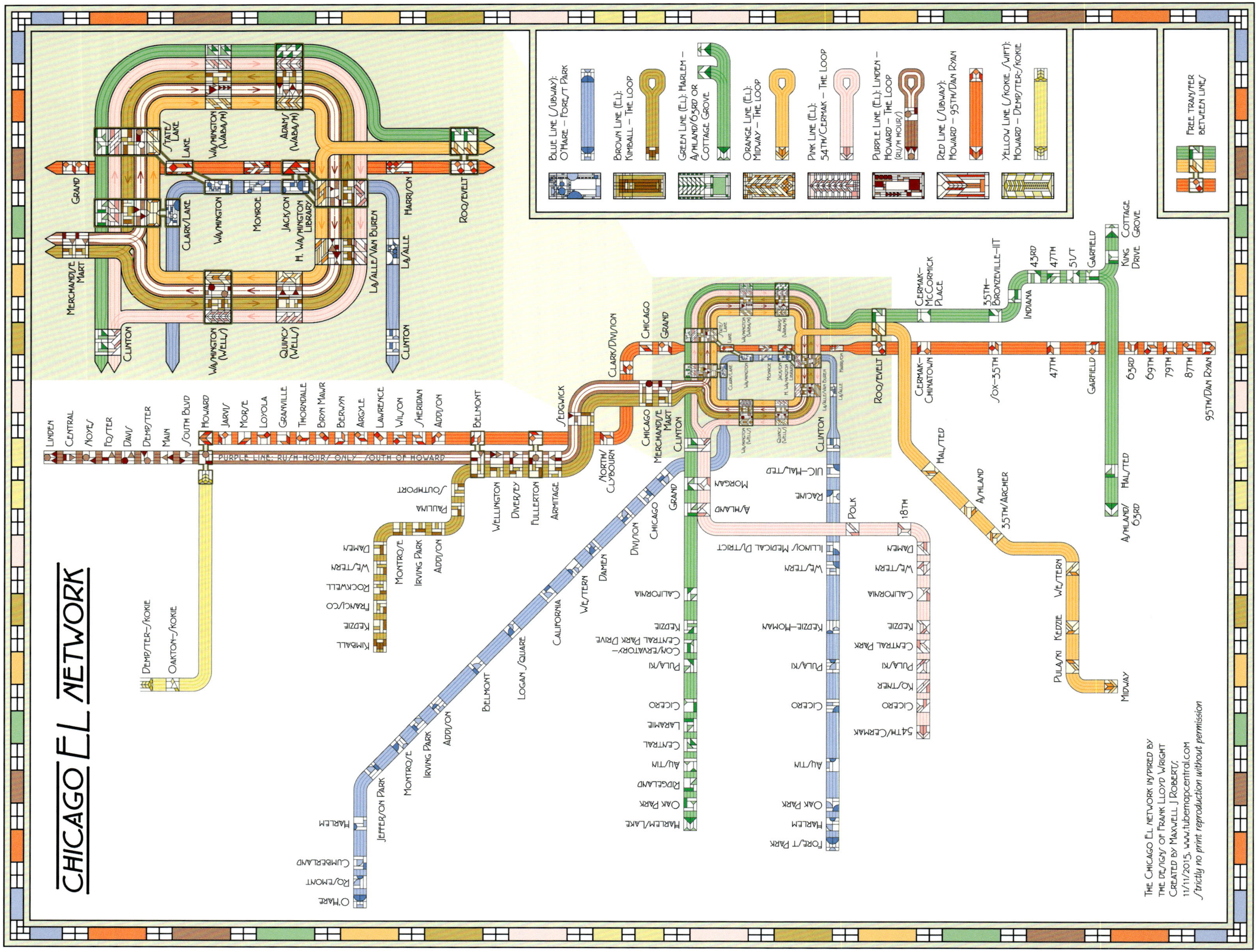
CHICAGO EL NETWORK
Blue Line (Subway): O'Hare – Forest Park
Brown Line (El): Kimball – The Loop
Green Line (El): Harlem – Ashland/63rd or Cottage Grove
Orange Line (El): Midway – The Loop
Pink Line (El): 54th/Cermak – The Loop
Purple Line (El): Linden – Howard – The Loop (rush hours)
Red Line (Subway): Howard – 95th/Dan Ryan
Yellow Line (Skokie Swift): Howard – Dempster-Skokie
Free transfer between lines
Purple Line: rush hours only south of Howard
Linden
Central
Noyes
Foster
Davis
Dempster
Main
South Blvd
Howard
Jarvis
Morse
Loyola
Granville
Thorndale
Bryn Mawr
Berwyn
Argyle
Lawrence
Wilson
Sheridan
Addison
Belmont
Dempster-Skokie
Oakton-Skokie
Southport
Paulina
Wellington
Diversey
Fullerton
Armitage
Sedgwick
North/Clybourn
Clark/Division
Chicago
Grand
Damen
Montrose
Irving Park
Addison
Western
Rockwell
Francisco
Kedzie
Kimball
Merchandise Mart
Clinton
State/Lake
Lake
Washington (Wabash)
Adams (Wabash)
Clark/Lake
Washington
Monroe
Jackson
H. Washington Library
LaSalle/Van Buren
LaSalle
Harrison
Roosevelt
Washington (Wells)
Quincy (Wells)
O'Hare
Rosemont
Cumberland
Harlem
Jefferson Park
Montrose
Irving Park
Addison
Belmont
Logan Square
California
Western
Damen
Division
Chicago
Grand
UIC-Halsted
Racine
Polk
Illinois Medical District
Western
Kedzie-Homan
Pulaski
Cicero
Austin
Oak Park
Harlem
Forest Park
Morgan
Ashland
California
Kedzie
Conservatory-Central Park Drive
Pulaski
Cicero
Laramie
Central
Austin
Ridgeland
Oak Park
Harlem/Lake
18th
Damen
Western
California
Kedzie
Central Park
Pulaski
Kostner
Cicero
54th/Cermak
Cermak-Chinatown
Sox-35th
47th
Garfield
63rd
69th
79th
87th
95th/Dan Ryan
Cermak-McCormick Place
35th-Bronzeville-IIT
Indiana
43rd
47th
51st
Garfield
King Drive
Cottage Grove
Halsted
Ashland/63rd
Halsted
Ashland
35th/Archer
Western
Kedzie
Pulaski
Midway
The Chicago El network inspired by the designs of Frank Lloyd Wright
Created by Maxwell J Roberts,
11/11/2015. www.tubemapcentral.com
Strictly no print reproduction without permission

《 Chicago El Map Inspired by Frank Lloyd Wright

Maxwell J. Roberts

The practical purpose of a map is to show the topography of a landscape so that people can navigate it effectively, but any map design can also have an aesthetic purpose. Even humble transit maps often include artistic flourishes. Stripped down schematics achieve a powerful simplicity, as in Henry Beck's London Underground map and Massimo Vignelli's New York City subway map, both of which are practical and beautiful. But if the goal of a map is to be more decorative than usable, then anything goes, and the result can achieve mesmerizing beauty.

I have always loved the stained glass of architect Frank Lloyd Wright's Prairie Houses; the way the patterns *seem* to repeat, yet never quite repeat, is sheer geometric bliss. Wright designed many houses in and around Chicago. The El (or L) transit network, with its robust grid, is a perfect setting for celebrating Wright's early style.

Planning and attention to detail are essential in a decorative transit map so that viewers will not see a jumble of stylistic clichés. The overall style and metaphors must be chosen carefully, and the cartographer must consider how the elements will fit together to function harmoniously. Eight Frank Lloyd Wright windows, each reenvisioned in two colors, are the inspirations for this map. The windows provide a graphic identity for the station tiles of the eight lines. Prototyping helped me identify how much intricacy the map needed and could withstand. I realized that each station tile had to be simple but unique, with strands along each line to bind them together. At joint stations, I combined and readjusted tiles in a painstaking process of trial and error to ensure that they did not clash with each other.

For transit map designers, battles with Chicago topography always end in defeat. I worked with the grid, allowing it to guide my map's structure and organization. The layout of the lines works well: if the decoration is stripped away, the shape of Chicago is revealed as both simplified and highlighted.

Creating this map required four years from initial test sketches to completion. I reached Version H before I was satisfied. The result is a map *inspired* by Frank Lloyd Wright, rather than *in his style*. The famously cantankerous architect would have been aghast at any attempt to impersonate him, but I hope he would forgive me after taking a closer look.

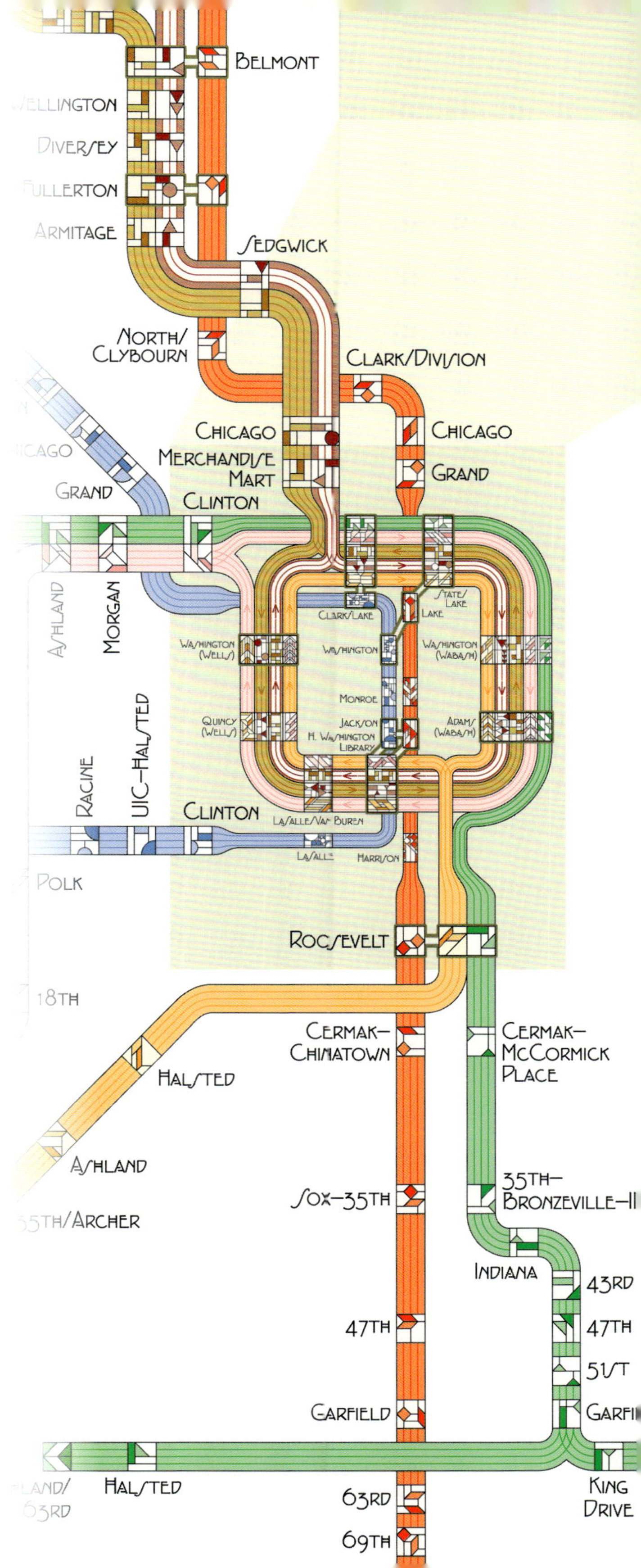

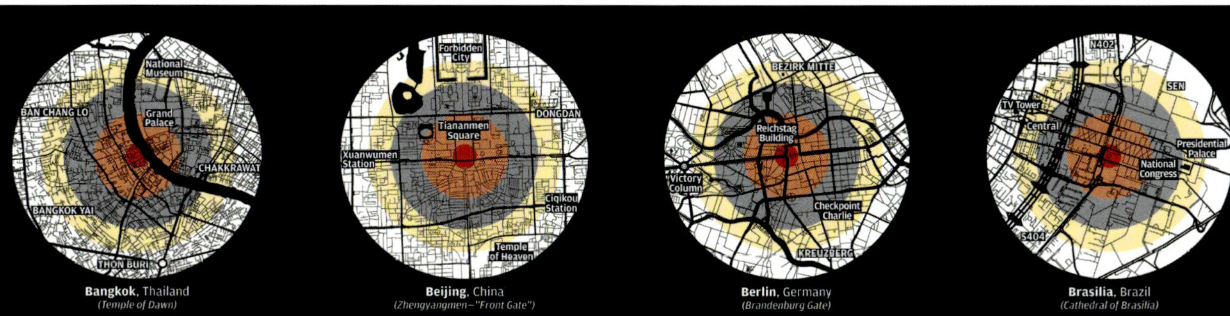

What if it was your city?

On August 6, 1945, the Japanese city of Hiroshima suffered the hitherto unknown devastation of a nuclear attack. On the orders of US President Harry Truman, a uranium bomb codenamed Little Boy was dropped at 8.15am. The US expected Japan to immediately capitulate, but the Japanese High Command mistakenly assumed the US only had one such bomb. At 11.02am on August 9, the spectacle of nuclear annihilation was repeated in Nagasaki with the dropping of Fat Man, a plutonium bomb. Five days later the Japanese surrendered unconditionally to Allied forces. With this, the second world war came to a close. Seventy years after the attacks, which killed a total of 200,000 people in Hiroshima and Nagasaki, the world is still not free from the threat of nuclear destruction, with nine countries possessing 19,000 nuclear weapons

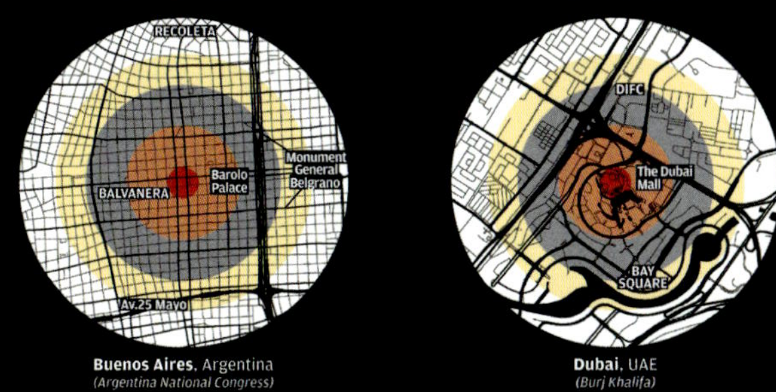

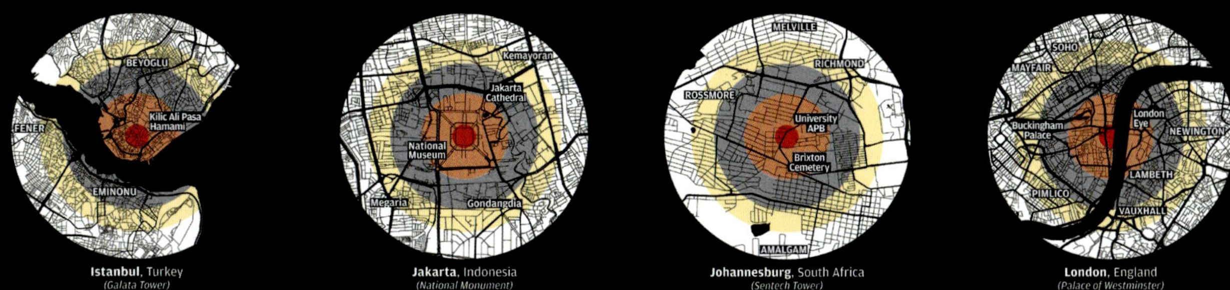

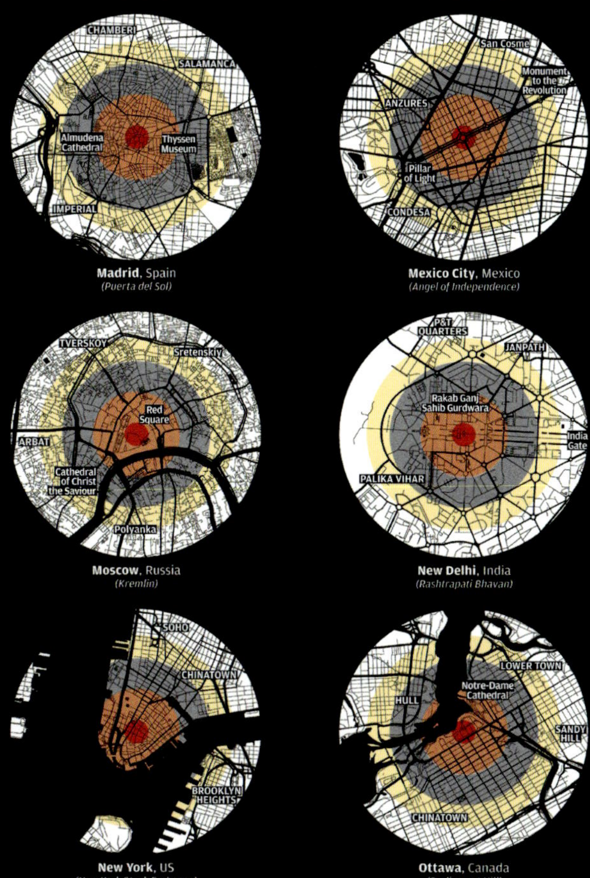

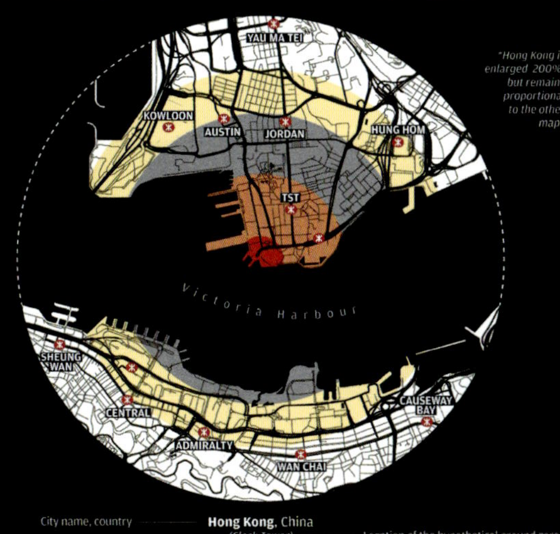

To illustrate the destructive power of these bombs, we analyse the impact the Fat Man bomb would have on modern cities around the world. We indicate the radiuses of immediate destruction upon detonation of our hypothetical explosion:

FIREBALL RADIUS
0.2km (0.13km²)
The nuclear fireball. The effect of its impact depends on height of detonation. If it touches the ground, the amount of radioactive fallout is significantly increased. Minimum burst height for negligible fallout is 180m

AIR BLAST RADIUS
0.76km (1.82km²)
Concrete buildings are severely damaged or demolished; fatalities approach 100%. Optimal height of burst to maximise this effect is 490m

RADIATION RADIUS
1.31km (5.42km²)
Mortality rates of between 50% and 90% are expected after exposure to radiation doses of 500rem. Death follows between several hours and several weeks

AIR BLAST RADIUS
1.72km (9.26km²)
Most residential buildings collapse, injuries are universal, fatalities are widespread. Optimal height of burst to maximise this effect is 0.85km

THERMAL RADIATION RADIUS
2.21km (15.4km²)
Third-degree burns extend throughout the layers of skin, and are often painless because they destroy the pain nerves. They can cause severe scarring or disability, and can require amputation

Note: The predictions draw on the detonation of plutonium bomb Fat Man (20kt) at Nagasaki, air burst at a height of 502m. The areas represented are approximate. We do not take into account subsequent radioactive fallout or a city's topography

Sources: Nukemap, The Making of the Atomic Bomb

SCMP Graphic: Alberto Lucas López

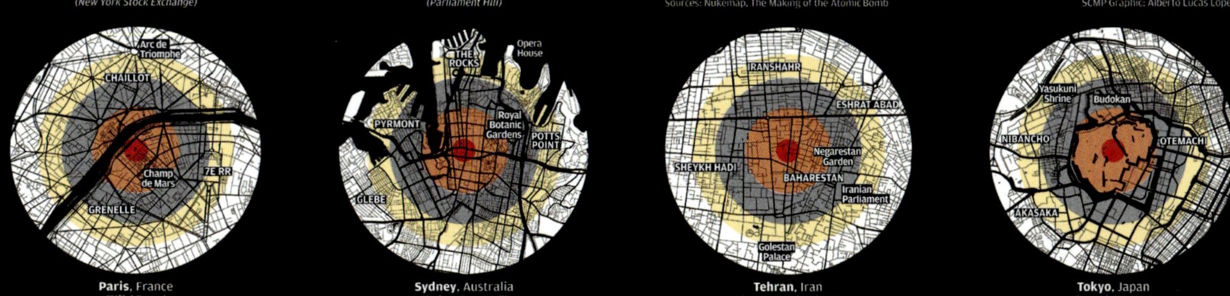

« What If It Was Your City?

Alberto Lucas López — South China Morning Post

In August 1945 the United States dropped two atomic bombs on Japan—"Little Boy" on Hiroshima and "Fat Man" on Nagasaki. I wanted to use the 70th anniversary of the events as an opportunity to explore the effects of the attacks and to tell the story in a way that readers could relate to personally. I decided the best way to do this was to visualize the explosions in contemporary cities across the globe.

This infographic was published in the *South China Morning Post*, a Hong Kong-based, English-language newspaper with a diverse readership. By focusing on neighborhoods obliterated in 20 cities in the home countries of many of the *Post's* readers, I knew that our audience could draw on personal experiences and memories to understand the magnitude of the horrors experienced decades ago in Japan.

I based the infographic on the more powerful Fat Man, transferring the bomb's impact from Nagasaki to the heart of each of the 20 cities. I indicated the extent of destruction based on the immediate detonation of an explosion, including the fireball, radiation, air blasts, and thermal radiation radii. And I chose a recognizable tourist destination as ground zero in each case: the Eiffel Tower in Paris, Temple of Dawn in Bangkok, and the Burj Khalifa in Dubai.

The map of each city is drawn to the same scale, except the map of my newspaper's home of Hong Kong, which is larger. In my research, I discovered that the number two was a recurring factor of importance, so I arranged the small maps into the shape of a numeral "2." The palette is somber and printed using a rich four-color black, with flashes of other colors to describe each radius.

The story told in the infographic highlights the power of nuclear weapons by showcasing the destruction they would inflict on cities readers know well. At the time I created this piece there were nine known nuclear powers in the world and 19,000 nuclear warheads. I wanted to show the possible results of unleashing Armageddon again if any of those weapons ended up in the wrong hands.

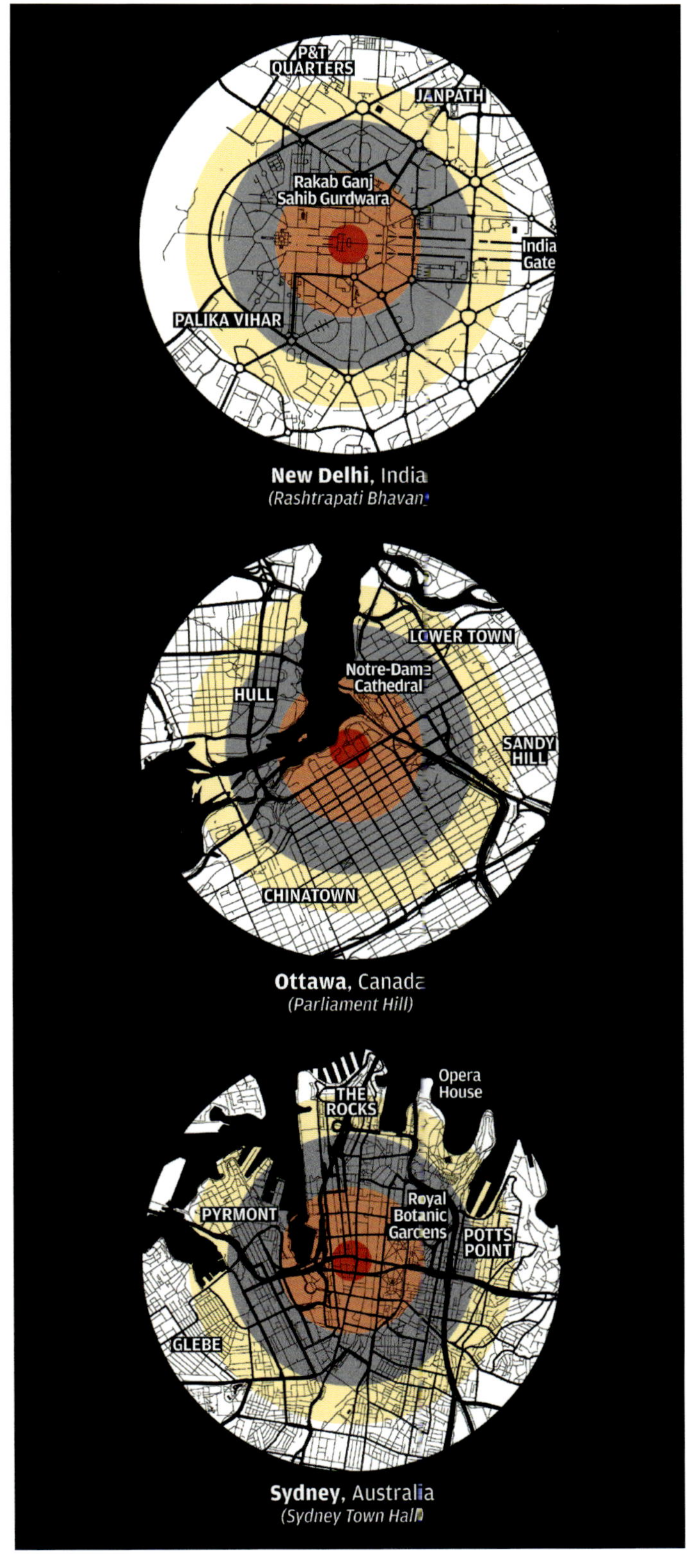

Cartography is an art of both place and time.

Sierra Nevada Snow Cover 2017 »

Nathaniel A. Douglass

The early months of 2017 kicked off one of the highest snow cover years for the Sierra Nevada in more than a decade. This map captures the dramatic increase in snow cover, while also portraying the breathtaking beauty of the mountains.

I grew up in the East Bay of San Francisco. After college, I moved to a small town in the foothills of the Sierra Nevada. There I found a new appreciation for this special landscape of deep emerald lakes, crystal blue rivers, and towering monoliths. I felt honored to live there.

Because the Sierra Nevada spans much of the California-Nevada border, this map needed a vertical layout and custom projection to accentuate the mountain range. I'm grateful to Aaron Taveras, Humboldt State University Geospatial Studies lecturer, for valuable input on the projection. Light blue graticules help orient the reader and provide a beautiful aesthetic addition. Satellite imagery collected between January 1 and March 30, 2017 provided the information needed to show the extent of snow cover. I created a false-color image, highlighting areas covered in snow, then extracted those areas and used them as a masking layer to manipulate and texturize the snow.

I used a lightened Swiss-style hillshade relief with isolated light blue shadowing on the snow, giving it a little more "frostiness." Ensuring that the snow cover had texture and dimension was crucial to the aesthetic of the map so that the snow would never appear to be a flat white blanket on top of tree cover. I achieved texture and dimension by adding a slight gradient to distinguish snow-covered areas at different elevations and added some embossing around the snow line.

Cartography is an art of both place and time. I hope that this map represents a unique snapshot of what the always-dynamic Sierra Nevada looked like during a brief moment in its long life. Many hours of love, joy, and even tears went into this map, so I hope viewers will handle it with care and have fun exploring a landscape near and dear to so many of us.

Learn more at ndcartography.com

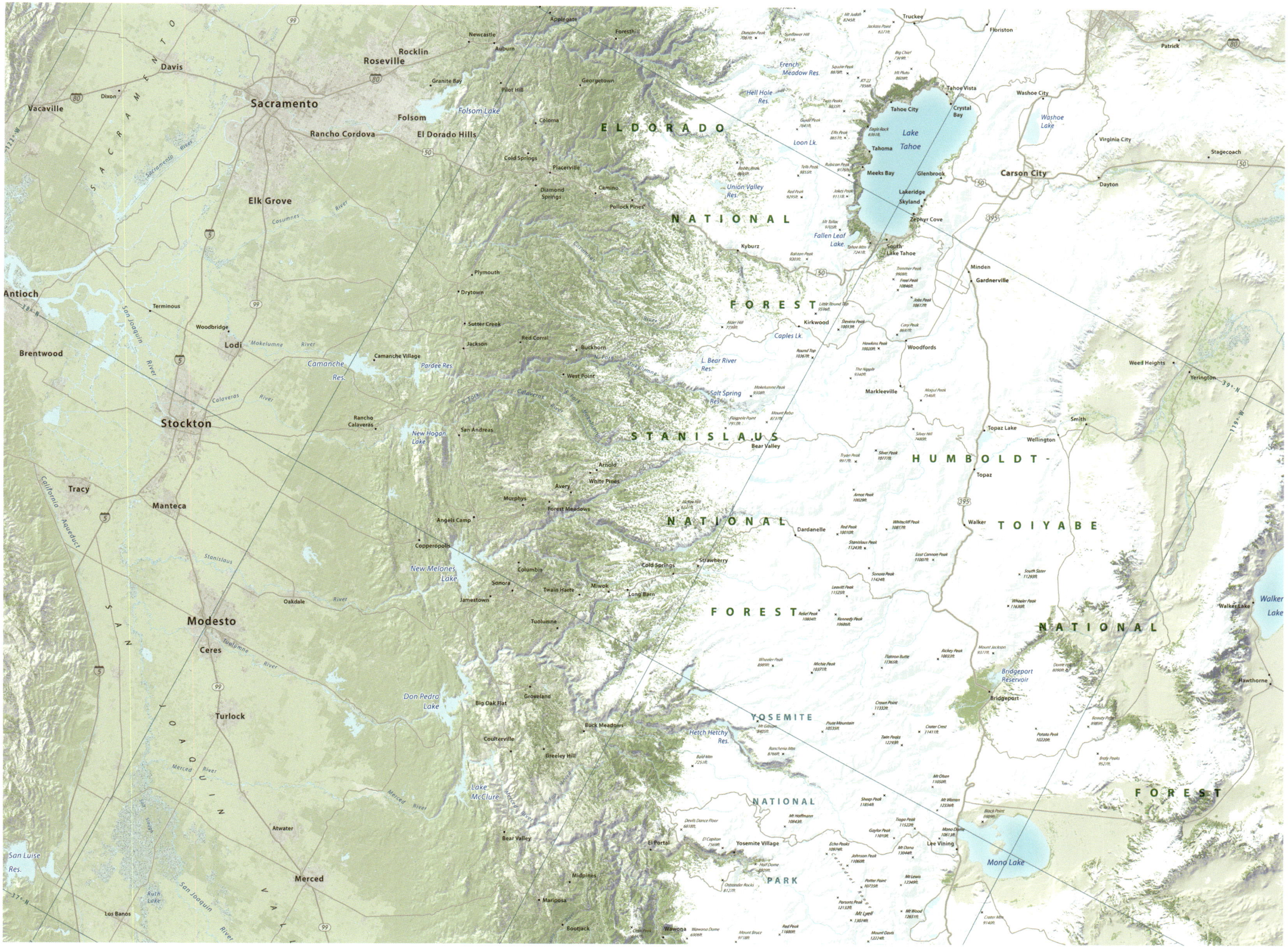

Sacramento
Roseville
Rocklin
Davis
Dixon
Vacaville
Folsom
Folsom Lake
Rancho Cordova
El Dorado Hills
Newcastle
Auburn
Applegate
Foresthill
Granite Bay
Pilot Hill
Georgetown
Coloma
Cold Springs
Placerville
Diamond Springs
Camino
Pollock Pines
Elk Grove
Cosumnes
River
Sacramento River
SACRAMENTO
ELDORADO
NATIONAL
FOREST
Antioch
Brentwood
Terminous
Woodbridge
Lodi
Mokelumne
River
San Joaquin River
Camanche Res.
Camanche Village
Pardee Res.
Plymouth
Drytown
Sutter Creek
Jackson
Red Corral
Buckhorn
West Point
Stockton
Calaveras
Rancho Calaveras
New Hogan Lake
San Andreas
Tracy
Manteca
California Aqueduct
Stanislaus
Copperopolis
Angels Camp
Murphys
Avery
Arnold
White Pines
Forest Meadows
New Melones Lake
Columbia
Sonora
Jamestown
Twain Harte
Mi-Wuk
Long Barn
Cold Springs
Strawberry
Oakdale
Modesto
Ceres
Tuolumne
River
Turlock
Don Pedro Lake
Big Oak Flat
Groveland
Buck Meadows
Coulterville
Greeley Hill
Merced
River
Lake McClure
Atwater
Merced
Bear Valley
Midpines
Mariposa
Bootjack
El Portal
Wawona
San Luise Res.
Los Banos
Ruth Lake
San Joaquin River
SAN JOAQUIN VALLEY
STANISLAUS
NATIONAL
FOREST
YOSEMITE
NATIONAL
PARK
Hetch Hetchy Res.
Yosemite Village
Kyburz
Union Valley Res.
Loon Lk.
Hell Hole Res.
French Meadow Res.
Truckee
Floriston
Tahoe Vista
Tahoe City
Crystal Bay
Lake Tahoe
Tahoma
Meeks Bay
Glenbrook
Lakeridge
Skyland
Zephyr Cove
Fallen Leaf Lake
South Lake Tahoe
Kirkwood
Caples Lk.
L. Bear River Res.
Salt Spring Res.
Bear Valley
Woodfords
Markleeville
Dardanelle
Washoe City
Washoe Lake
Virginia City
Carson City
Stagecoach
Dayton
Patrick
Minden
Gardnerville
Weed Heights
Yerington
Smith
Topaz Lake
Wellington
Topaz
Walker
HUMBOLDT-
TOIYABE
NATIONAL
FOREST
Bridgeport Reservoir
Bridgeport
Walker Lake
Walker Lake
Hawthorne
Lee Vining
Mono Lake

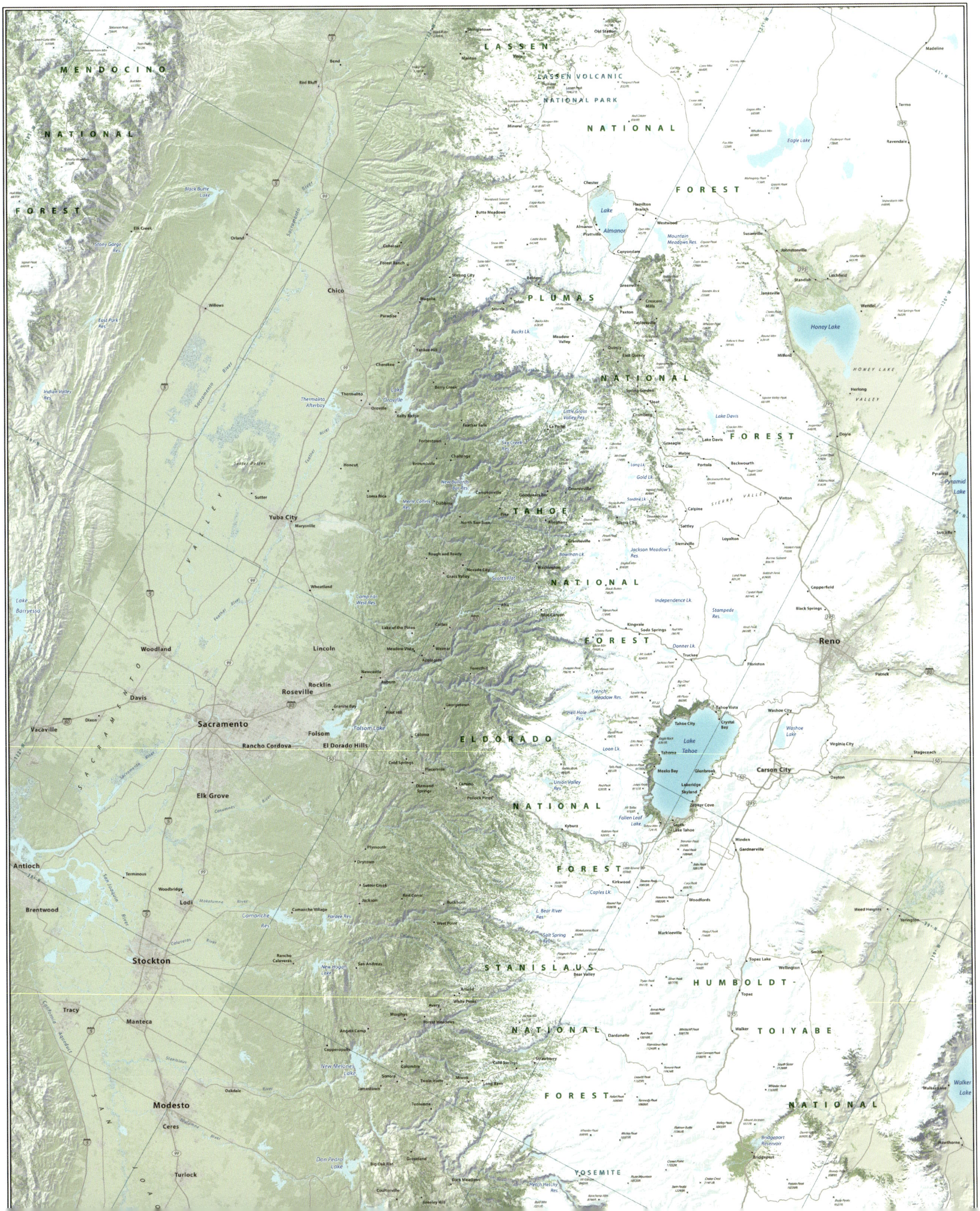
MENDOCINO
NATIONAL
FOREST
LASSEN
LASSEN VOLCANIC
NATIONAL PARK
NATIONAL
FOREST
PLUMAS
NATIONAL
FOREST
TAHOE
NATIONAL
FOREST
ELDORADO
NATIONAL
FOREST
STANISLAUS
NATIONAL
FOREST
HUMBOLDT-
TOIYABE
NATIONAL
YOSEMITE
SACRAMENTO
VALLEY
SAN JOA
Honey Lake
HONEY LAKE
VALLEY
Eagle Lake
Lake Almanor
Lake Tahoe
Pyramid Lake
Walker Lake
Washoe Lake
Lake Davis
Bucks Lk.
Stampede Res.
Independence Lk.
Donner Lk.
Jackson Meadows Res.
Loon Lk.
Union Valley Res.
Fallen Leaf Lake
Caples Lk.
Folsom Lake
Lake Oroville
Thermalito Afterbay
Camp Far West Res.
New Melones Lake
New Hogan Lake
Don Pedro Lake
Camanche Res.
Pardee Res.
Black Butte Lake
Stony Gorge Res.
East Park Res.
Indian Valley Res.
Lake Berryessa
Hetch Hetchy Res.
Bridgeport Reservoir
Sacramento River
Feather River
San Joaquin River
California Aqueduct
Sutter Buttes
Reno
Carson City
Sacramento
Stockton
Modesto
Chico
Yuba City
Marysville
Roseville
Rocklin
Lincoln
Folsom
El Dorado Hills
Rancho Cordova
Elk Grove
Lodi
Woodland
Davis
Vacaville
Dixon
Antioch
Brentwood
Tracy
Manteca
Ceres
Turlock
Oroville
Red Bluff
Orland
Willows
Susanville
Truckee
Tahoe City
South Lake Tahoe
Minden
Gardnerville
Quincy
Auburn
Placerville
Jackson
Sonora
Wheatland
Sutter
Paradise
Bridgeport
Hawthorne
Madeline
Termo
Ravendale
Chester
Westwood
Portola
Loyalton
Sierraville
Kingvale
Soda Springs
Grass Valley
Nevada City
Colfax
Markleeville
Woodfords
Kirkwood
Kyburz
Topaz
Walker
Wellington
Smith
Yerington
Dayton
Virginia City
Stagecoach
Patrick
Black Springs
Copperfield
Doyle
Herlong
Milford
Janesville
Standish
Litchfield
Wendel
Johnstonville
Sutcliffe
Pyramid
Vinton
Calpine
Sattley
Sierra City
Downieville
Camptonville
Dobbins
Challenge
Brownsville
Forbestown
Feather Falls
Berry Creek
Cherokee
Thermalito
Honcut
Loma Rica
Penn Valley
Rough and Ready
Washington
Alta
Meadow Vista
Applegate
Newcastle
Granite Bay
Cool
Coloma
Georgetown
Foresthill
Camino
Pollock Pines
Diamond Springs
Cold Springs
Plymouth
Drytown
Sutter Creek
Ione
Camanche Village
Rancho Calaveras
San Andreas
Arnold
Murphys
Angels Camp
Copperopolis
Columbia
Jamestown
Twain Harte
Long Barn
Strawberry
Tuolumne
Groveland
Big Oak Flat
Coulterville
Oakdale
Terminous
Woodbridge
Elk Creek
Bend
Mineral
Manton
Shingletown
Old Station
Greenville
Crescent Mills
Taylorsville
Meadow Valley
East Quincy
Spring Garden
Graeagle
Clio
Mohawk
Beckwourth
Tahoe Vista
Crystal Bay
Incline Village
Glenbrook
Zephyr Cove
Stateline
Meeks Bay
Homewood
Tahoma
Washoe City
Pleasant Valley
Bear Valley
Dardanelle
Cold Springs
Pinecrest
Mi-Wuk Village

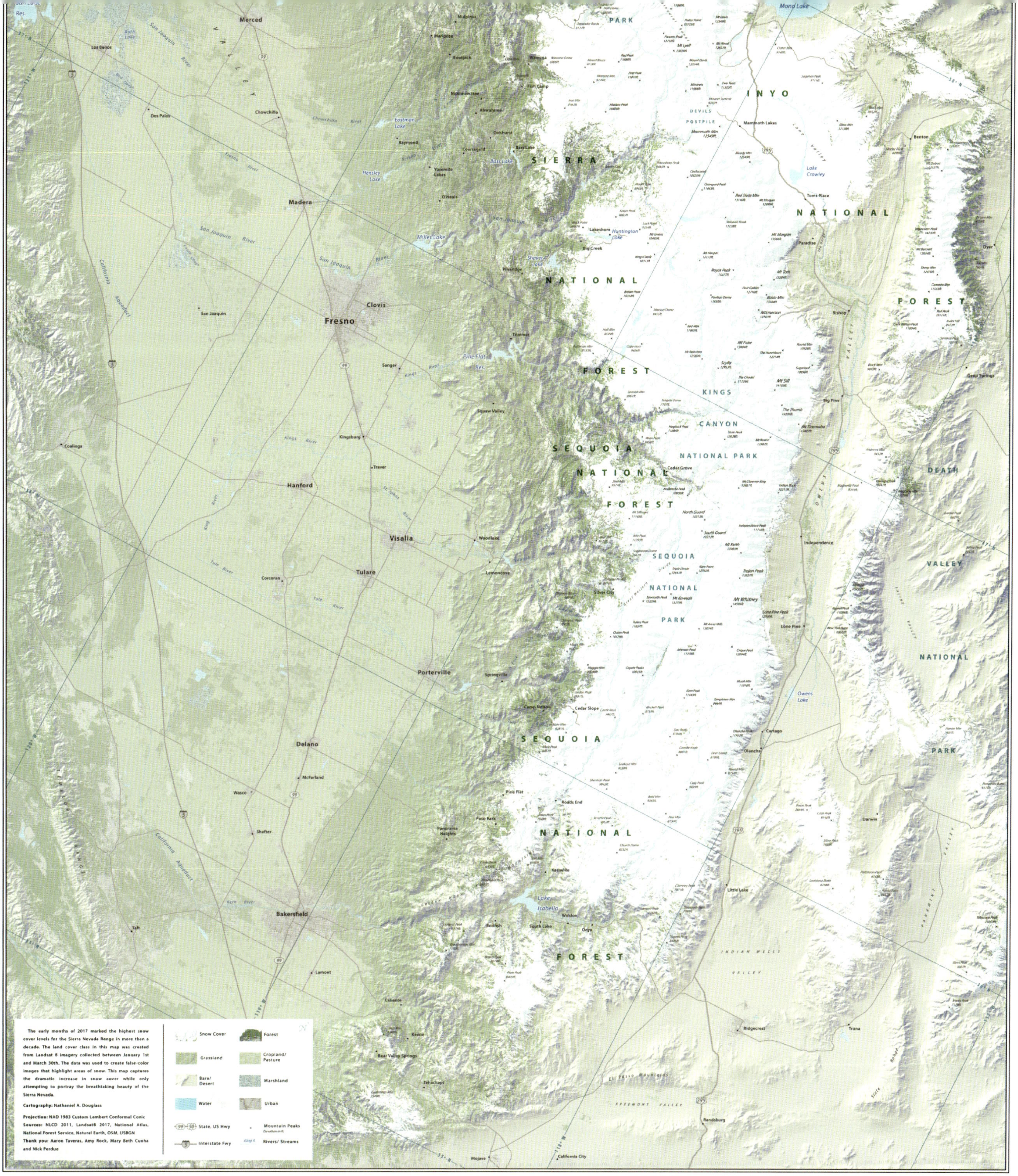

SIERRA NEVADA

SNOW COVER 2017

0 100 Mi

0 100 Km

1 : 700,000

A Visual History of U.S. State Boundaries »

Joshua Korenblat, Karl Phillips, Jeff Osborn, and Nathaniel Pearlman — Graphicacy

Peculiar shapes are everywhere on the map of the United States. Consider Connecticut's Southwick Jog, Delaware's Newcastle Arc, Missouri's bootheel, and Idaho's stovepipe. Our poster, *A Visual History of U.S. State Boundaries*, encourages an up-close reading of the natural and human events that gave the states their distinctive shapes.

Our team worked with historian and cartographer Karl Phillips to transform more than 800 records into a layered visualization. Before we embarked on the design, we wondered if the visualization might become unreadable with so many records. We soon discovered that unlike other types of data visualization, geographic maps could communicate far more than five layers of information. We counted one crisp corner of a classic Swiss mountain map by Eduard Imhof and discovered there were 14 layers of data presented in points, lines, planes, and color. Sometimes, as many designers have noted, we don't simplify; we clarify.

Our map represents present-day boundaries as strokes, proposed and annexed boundaries as fills, and historical or abandoned boundaries as dashed lines. Colors mark boundary types: multicolored lines indicate more than one category for a single state boundary; shades of black—inspired by the work of Paula Scher—blanket the topographical map, emphasizing the colorful array of lines. These shades also suggest former international territories that are now part of the United States.

The shapes of the lines also invite closer inspection. Squiggly lines indicate natural state boundaries, including mountains and shorelines. Straight lines, zigzags, jogs, circles, the bootheel, and the stovepipe all inspire questions about the people who drew them. Surveying errors inform several state shapes today, including the Walker Line dividing Kentucky and Tennessee.

A close reading of the map leads viewers to design surprises. In the first timeline entry along the bottom of the poster, we see that the original colonial charter lines spanned sea to sea. Some scholars suggest that this pattern inspired the design of the thirteen stripes on the U.S. flag. While Thomas Jefferson imagined the human-made boundaries of states between the Appalachians and the Mississippi River as rectangular and the states as equal in relative area, this vision only took form later, as settlers moved farther west.

Most posters present hierarchies of text and images to communicate a simple, bold message. We want readers to approach posters like this one as if reading a book with a fascinating story to tell.

A VISUAL HISTORY OF THE
US STATE BOUNDARIES

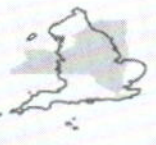

MAJOR PRINCIPLES GOVERNING STATE SIZE

British monarchs had little concern with establishing equity among the colonies, but even the largest colonies were eventually reduced in size. New York was gradually reduced to an area comparable to England itself, which was considered to be the ideal size of a political state. After the American Revolution, the US struggled with the issue of state size equity. Various proposals to consolidate the smaller states were met with resistance. When attempting to carve the western territories into new units, the Founding Fathers sought to limit the new states in both size and number. Frontier states were granted about 50,000 square miles, roughly the size of the largest of the original states. By 1787, the number of new states would be limited to twelve so as to protect the political power of the thirteen original states in Congress. When the United States acquired the Louisiana Purchase in 1803, it became impossible to limit both the size and number of new states. A choice had to be made: create much larger states west of the Mississippi River, or continue adding new states comparable in size to the larger states such as New York, Georgia and Kentucky. Thomas Jefferson, the President, counseled his peers that the Union was strong enough to add as many states as needed to preserve the prevailing republican principles of the period. Louisiana, Missouri, Arkansas and Iowa were admitted with territories comparable in size to neighboring states east of the Mississippi River. The admission of Texas turned the tide in favor of creating large Western states. The 1845 Joint Congressional Resolution, providing for the annexation

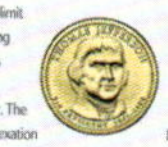

and admission of Texas, included a provision allowing five total states to be created from the Texas Republic. This seemed to confirm the established principle of size equity among the frontier states. However, other than the reduction of Texas in 1850, there has never been another successful effort to reduce Texas. Congress sought to preserve the political power of the eastern states by admitting large Western states. By 1859, a new standard of 100,000 square miles was adopted for Western states, beginning with the admission of Oregon that year. The Plains states north of Texas each have territories of about 75,000 square miles, representing a transition zone between the humid eastern states and the larger arid Western states. California and Montana are well over 100,000 square miles, but they are offset by states such as Utah, Washington and Idaho, which are well under the Western standard.

STATES WEST OF THE MISSISSIPPI RIVER

Louisiana 1812 — Known as the Florida parishes, the area between the Mississippi and Pearl Rivers was once part of the Spanish colony of West Florida. It became part of Orleans Territory in 1804 and was included in **Louisiana** (29) upon statehood in 1812.

Missouri 1821 — The Osage Line (30) marked the western limit of white settlement in the Missouri Valley in 1816. It was selected as Missouri's western boundary upon statehood in 1821. The Platte Purchase was acquired in 1837, giving Missouri its northwestern triangle.

Arkansas 1836 — A 980-square-mile notch in the northern boundary of Arkansas, known as the Missouri Bootheel, (31) was included in Missouri upon its statehood. This anomaly occurred because the town of Caruthersville wanted to be included in Missouri rather than Arkansas.

Texas 1845 — In 1850, Texas sold a part of its territory to the US. This area became half of present-day New Mexico, a third of Colorado, and small portions of Kansas, Oklahoma and Wyoming. It retained the **Western Panhandle** (32) for use as a railroad corridor.

Iowa 1846 — The object of the Honey War, a territorial dispute between Iowa and Missouri, the **Sullivan Line** (33) was first surveyed in 1816 to mark a key Indian cession. Iowa claimed it was too far north. Missouri took the opposite view. The Supreme Court of the United States upheld its validity in 1849.

California 1850 — California's eastern boundary was drawn so that the Golden State would include all of the gold-rich **Sierra Nevada Range**. (34) The Weller Line (35) was selected as the southern boundary to include the port of San Diego.

- Minnesota 1858
- Oregon 1859
- Kansas 1861
- Nevada 1864
- Nebraska 1867
- Colorado 1876
- North Dakota 1889
- South Dakota 1889
- Montana 1889
- Washington 1889
- Idaho 1890
- Wyoming 1890
- Utah 1896
- Oklahoma 1907
- New Mexico 1912
- Arizona 1912
- Alaska 1959
- Hawaii 1959

STATES EAST OF THE MISSISSIPPI RIVER

Delaware 1787 — The New Castle Arc (01) forms the boundary between Delaware and Pennsylvania. It has a radius of twelve miles, with its origin at the dome of the old county courthouse in New Castle, a town that was Delaware's colonial capital.

- Pennsylvania 1787
- New Jersey 1787
- Georgia 1788
- Connecticut 1788
- Massachusetts 1788
- Maryland 1788
- South Carolina 1788
- New Hampshire 1788
- Virginia 1788
- New York 1788
- North Carolina 1789
- Rhode Island 1790
- Vermont 1791
- Kentucky 1792
- Tennessee 1796
- Ohio 1803
- Indiana 1816
- Mississippi 1817
- Illinois 1818
- Alabama 1819
- Maine 1820
- Michigan 1837
- Florida 1845
- Wisconsin 1848
- West Virginia 1863

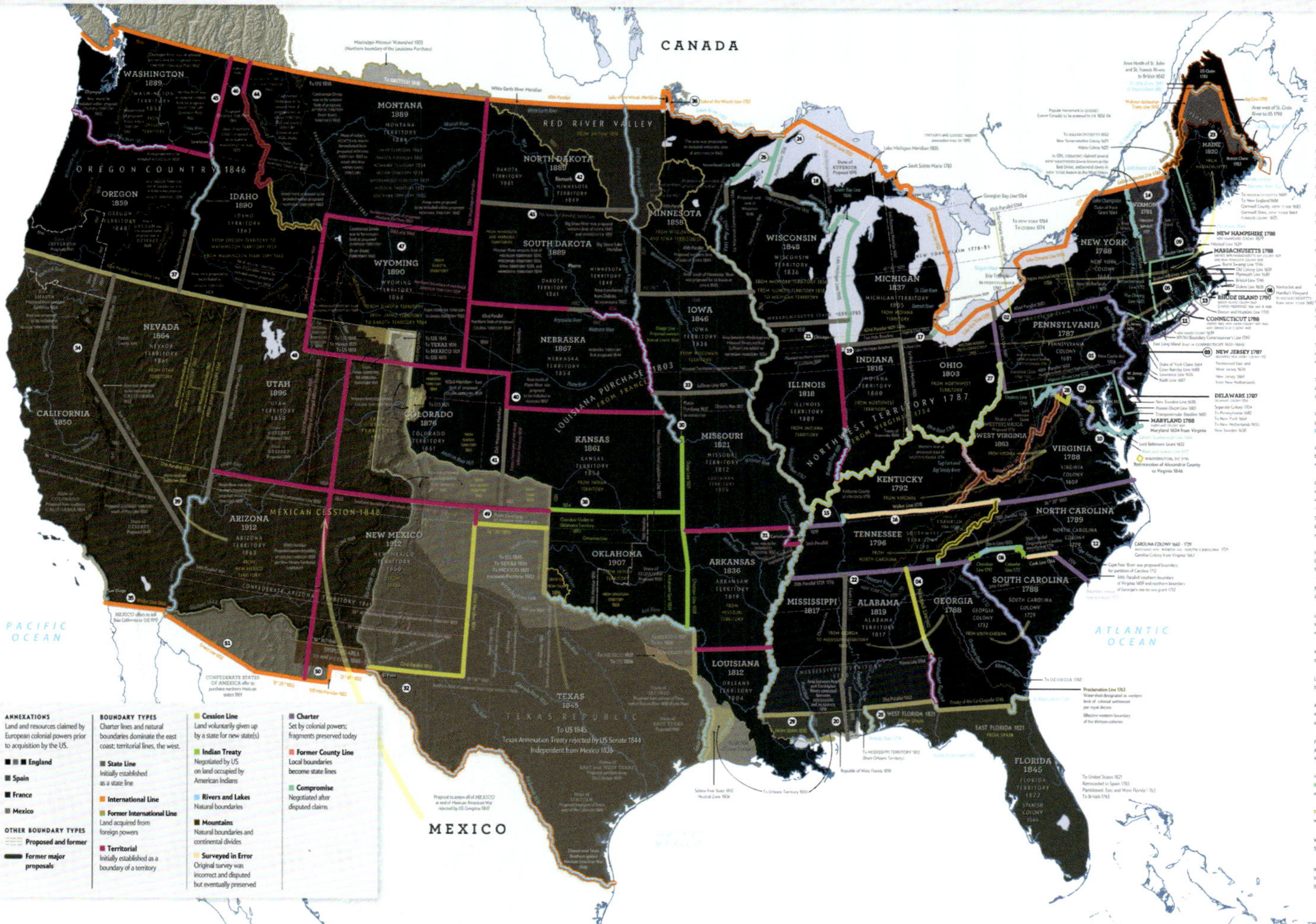

BOUNDARY TIMELINE

COLONIAL

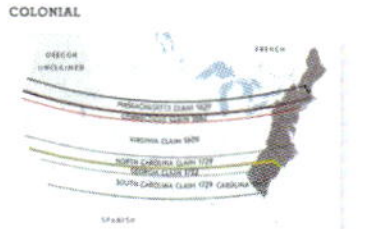

Each of the original thirteen colonies based their boundaries on various colonial grants and charters. In most cases, their claims conflicted with those of neighboring colonies and led to numerous territorial disputes. These claims were mostly resolved by the time of the American Revolution. However, some claims continued to fester well into the 19th century. Several colonies were chartered with generous sea-to-sea grants extending across the continent. These land claims instilled an early sense of ownership of the modern territory of the United States. Their visual pattern may have been an inspiration for the design of the American flag, with its thirteen stripes. Colonial sea-to-sea grants have also been associated with the 19th century doctrine of Manifest Destiny, which provided the US with a philosophical and even theological justification for acquiring territory in the West, displacing foreign powers and American Indians.

1776 INDEPENDENCE

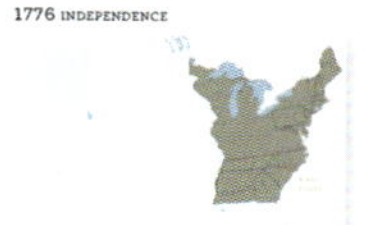

Perhaps more than any other issue, slavery profoundly affected the size, number, and shape of the states. Slavery was legal in each of the thirteen original states at the time of independence. Efforts by Thomas Jefferson to abolish slavery in the Declaration of Independence were overruled by southern delegates to the Continental Congress. Having long depended upon slavery on southern tobacco plantations, southern political leadership resisted any attempt to force abolition on the states. It was decided that the abolition of slavery should be left to the individual states on a voluntary basis. Vermont abolished slavery when it declared its statehood in 1777. Pennsylvania was the first of the original states to abolish slavery in 1780.

1798 FEDERAL ERA

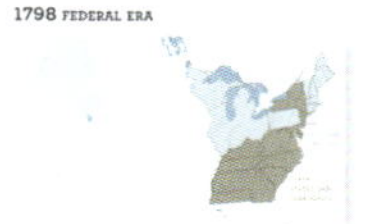

By 1784, slavery had been abolished in each of the New England states and Pennsylvania. In 1787, Congress abolished slavery in the Northwest Territory and from any future states created within the territory. Vermont entered the Union in 1791 as a free state. Kentucky entered the Union in 1792 and Tennessee in 1796, both as slave states. Mississippi Territory was organized in 1798 as a slave territory. Slavery was starting to wane, even in the South, and most Americans expected the remaining states to abolish slavery in the near future. The invention of the cotton gin in 1793 reversed that trend. Cotton was now a lucrative crop that required cheap intensive labor to plant and harvest. By 1810, slavery was an intrinsic part of the Southern economy. Southern resistance to abolition hardened and divided the nation along sectional lines. Cultivation of cotton spread across the South, and slavery spread across the cotton belt.

1822 MISSOURI COMPROMISE

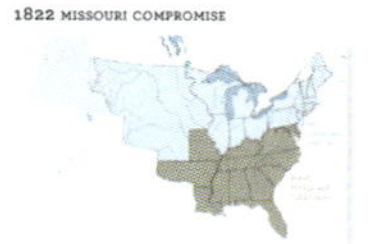

New York abolished slavery in 1799. Ohio entered the Union as a free state in 1803. New Jersey became the last northern state to abolish slavery in 1804. For the first time in US history, free states outnumbered slave states. Louisiana, acquired in 1803 as part of the Louisiana Purchase, entered the Union in 1812 as a slave state, equalizing the balance of free and slave states at nine each. States then entered the Union in pairs in order to maintain the balance of free and slave states: Indiana (free, 1816) and Mississippi (slave, 1817); Illinois (free, 1818) and Alabama (slave, 1819). The Missouri Compromise admitted Maine into the Union as a free state in 1820 to balance the admission of Missouri in 1821 as a slave state. Slavery was prohibited in the unorganized remnant of Missouri Territory north of 36° 30'. Michigan Territory, organized in 1805, remained free. Arkansaw Territory and Florida Territory, organized in 1819 and 1822 respectively, remained slave territories.

1854 KANSAS–NEBRASKA ACT

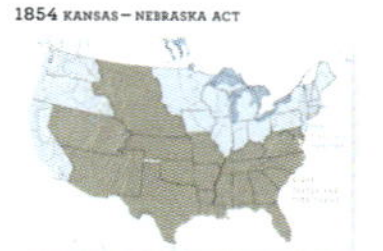

Following the admission of Arkansas (slave, 1836) and Michigan (free, 1837), Florida and Texas entered the Union as slave states in 1845, giving the South a majority in the Senate. In exchange for northern support for Florida's admission, southern congressmen admitted Iowa as a free state, which entered the union in 1846. Subsequently, ratification of the Treaty of Guadalupe-Hidalgo, which ended the Mexican-American War, required a two-thirds majority in the Senate. This forced southern congressmen to admit Wisconsin (free, 1848), which restored a balance. Texas was annexed in 1845 and admitted as a slave state, but as a compromise, in 1848, the Oregon Territory was organized as free. However, slavery was permitted in the Utah and New Mexico Territories. The Kansas-Nebraska Act overturned the Missouri Compromise by allowing slavery in two Great Plains territories. California was admitted as a free state, but its two senators split on the slavery question.

1861 CIVIL WAR

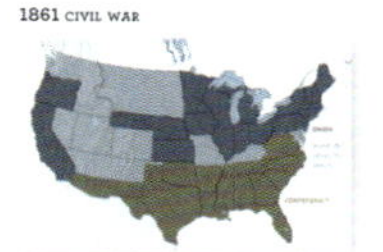

The Dred Scott decision by the Supreme Court opened slavery to all the territories in 1857. This decision, combined with anger over the Kansas-Nebraska Act, galvanized the North against slavery in the West. Minnesota entered the Union as a free state in 1858 with southerners expecting Kansas to be a slave state. However, Kansas voters rejected slavery in their constitution. Oregon was admitted as a free state in 1859. The census of 1860 indicated that the North would soon hold a House majority. After Abraham Lincoln's election, abolitionists would control Congress and the executive branch. Convinced slavery would be abolished, the South seceded from the United States and formed the Confederate States of America, precipitating the Civil War. Now, free states dominated Congress. Colorado, Dakota and Nevada Territories were organized as slave territories. Kansas entered the Union as a free state in 1861. Nebraska was the only western territory to voluntarily abolish slavery. The border slave states of Delaware, Maryland, Kentucky and Missouri remained loyal to the Union.

1865 RECONSTRUCTION

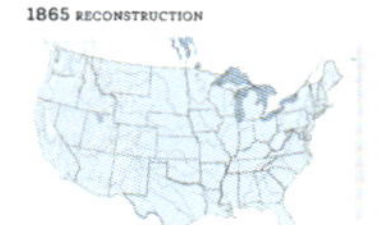

Reconstruction of the South started once the war ended in 1865. Prior to 1865, West Virginia entered the Union as a slave state in 1863 with provisions to gradually abolish slavery, and Nevada entered as a free state in 1864. In 1865, the Thirteenth Amendment abolished slavery. Confederate states were only readmitted into the Union upon ratifying the amendment. Nebraska entered in 1867 with Republican support; Colorado in 1876, entered with Democratic support. But in the 1876 presidential election, the Democrats were dismayed when Colorado voted for Republican candidate Rutherford B. Hayes; they then opposed statehood for any pro-Republican territories. Republicans won the House in 1889 and sped western statehood: North and South Dakota, Montana, Washington, Idaho and Wyoming (1889-90), and Utah in 1896. The Sequoyah statehood movement prompted Congress to admit Oklahoma in 1907, combining the Oklahoma and Indian Territories to avoid creating a state with an American Indian majority. New Mexico and Arizona (1912) completing the contiguous United States.

1959 ALASKA AND HAWAII

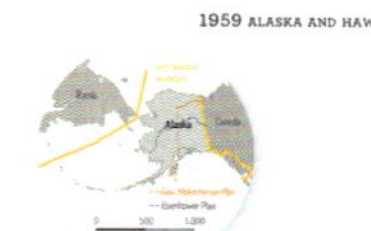

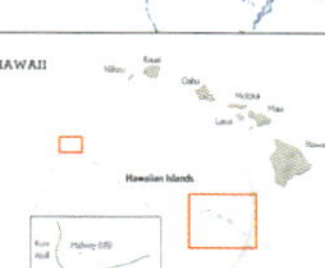

Discovered by Danish explorer Vitus Bering in 1741, Alaska was claimed and colonized by Russia. The eastern boundary was defined by the Anglo-Russo Treaty of 1825, which follows the 141st Parallel southward from the Arctic Ocean to Mount St Elias. From there it parallels the coast to the Dixon Entrance at latitude 54° 40' N. The US purchased Alaska in 1867. Its Bering Sea boundary runs roughly parallel to the International Date Line. Alaska was designated a District in 1884 and organized as a Territory in 1912. Proposals to divide the Territory in 1954 (Gov. Heintzleman Plan), and 1955 (President Eisenhower Plan) were rejected by Alaskans. Statehood was achieved in 1959.

The Sandwich Islands were discovered by British sea captain James Cook in 1778 on his Pacific voyage of discovery. They were unified into the Kingdom of Hawaii by King Kamehameha the Great in 1795. Successive rulers expanded the territory of the kingdom westward, eventually claiming the Kure Atoll by 1886. Neighboring Midway Islands were claimed by the United States in 1859 and were never officially part of Hawaii. The monarchy was overthrown by expatriate Americans in 1893 and a republic was declared in 1894. The US annexed Hawaii in 1898 as part of an expansion across the Pacific. Organized as a territory in 1900, Hawaii achieved statehood in 1959.

PUERTO RICO AND VIRGIN ISLANDS

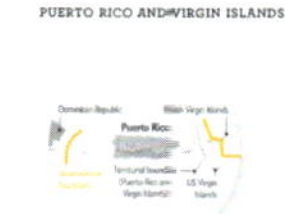

These islands were the first part of the United States to be colonized by Europeans. Both were discovered by Columbus on his first voyage in 1492. Spanish colonization of Puerto Rico began in 1508. By 1625, the Dutch began to colonize the Virgin Islands. Denmark gained control of the Virgin Islands in 1666 and named them the Danish West Indies. The United States acquired Puerto Rico in 1898 at the conclusion of the Spanish-American War. It was organized as a Territory in 1900 and became a Commonwealth in 1952. The Danish West Indies were purchased in 1917 to prevent Germany from acquiring them during World War I. They were officially renamed the United States Virgin Islands and organized as a Territory in 1954. In 2012, 800,000 Puerto Rican citizens voted in favor of statehood – 61% of those who cast ballots on the issue.

TIMEPLOTS
information graphic products

Visit **www.timeplots.com** for sources and more information.

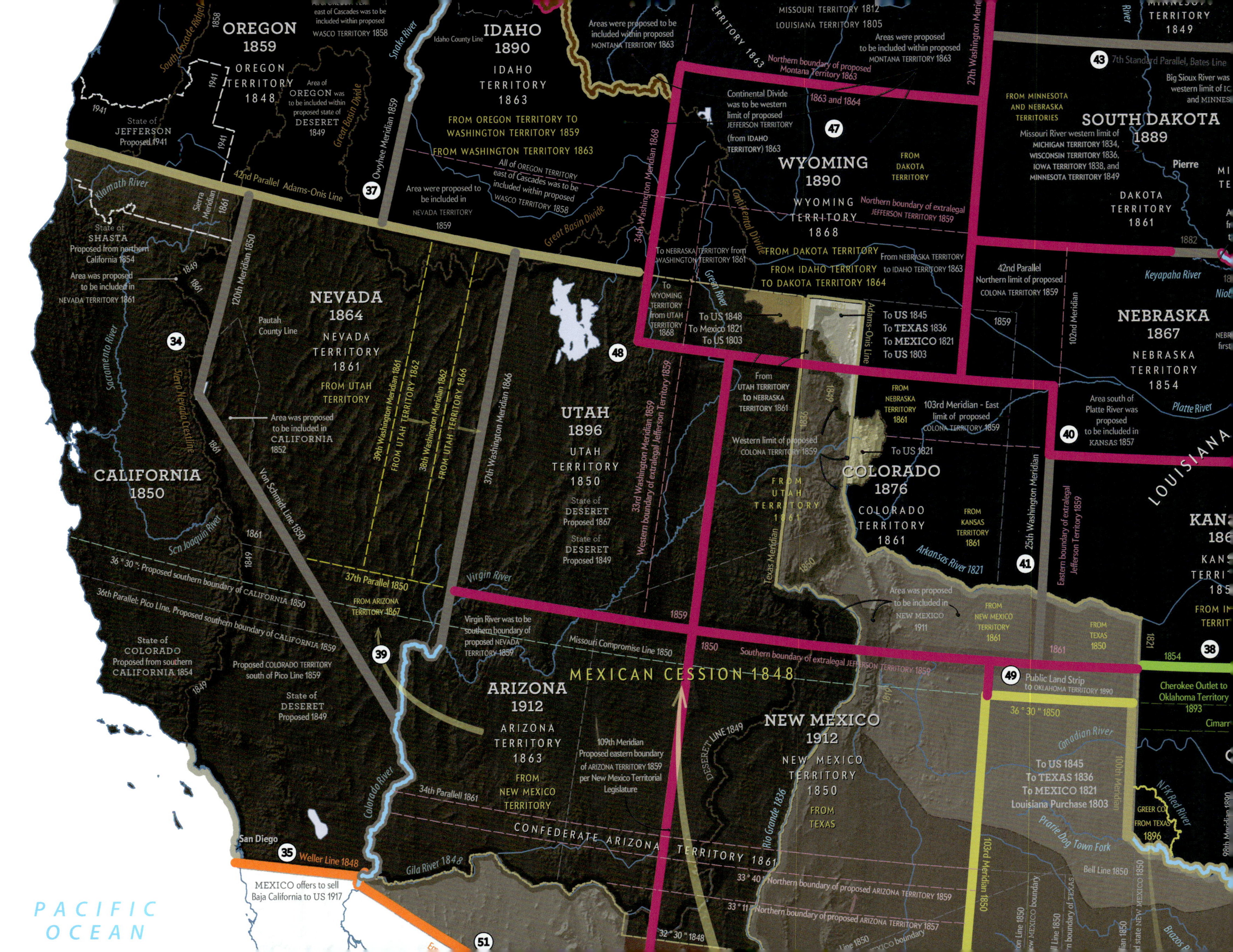
OREGON
1859
OREGON TERRITORY 1848
Area of OREGON was to be included within proposed state of DESERET 1849
State of JEFFERSON Proposed 1941
South Cascade Ridge
Great Basin Divide
east of Cascades was to be included within proposed WASCO TERRITORY 1858
Snake River
IDAHO
1890
IDAHO TERRITORY 1863
Idaho County Line
FROM OREGON TERRITORY TO WASHINGTON TERRITORY 1859
FROM WASHINGTON TERRITORY 1863
All of OREGON TERRITORY east of Cascades was to be included within proposed WASCO TERRITORY 1858
Area were proposed to be included in NEVADA TERRITORY 1859
Owyhee Meridian 1859
42nd Parallel Adams-Onis Line
Great Basin Divide
Areas were proposed to be included within proposed MONTANA TERRITORY 1863
TERRITORY 1863
MISSOURI TERRITORY 1812
LOUISIANA TERRITORY 1805
Areas were proposed to be included within proposed MONTANA TERRITORY 1863
Northern boundary of proposed Montana Territory 1863
27th Washington Meridian
MINNESOTA TERRITORY 1849
7th Standard Parallel, Bates Line
Big Sioux River was western limit of IO and MINNESOTA
Continental Divide was to be western limit of proposed JEFFERSON TERRITORY (from IDAHO TERRITORY) 1863
1863 and 1864
WYOMING
1890
WYOMING TERRITORY 1868
FROM DAKOTA TERRITORY
Northern boundary of extralegal JEFFERSON TERRITORY 1859
34th Washington Meridian 1868
Continental Divide
To NEBRASKA TERRITORY from WASHINGTON TERRITORY 1861
FROM DAKOTA TERRITORY
FROM IDAHO TERRITORY TO DAKOTA TERRITORY 1864
From NEBRASKA TERRITORY to IDAHO TERRITORY 1863
FROM MINNESOTA AND NEBRASKA TERRITORIES
SOUTH DAKOTA
1889
Missouri River western limit of MICHIGAN TERRITORY 1834, WISCONSIN TERRITORY 1836, IOWA TERRITORY 1838, and MINNESOTA TERRITORY 1849
Pierre
DAKOTA TERRITORY 1861
1882
Klamath River
Sierra Meridian 1861
State of SHASTA Proposed from northern California 1854
Area was proposed to be included in NEVADA TERRITORY 1861
120th Meridian 1850
Sacramento River
Sierra Nevada Crestline
NEVADA
1864
NEVADA TERRITORY 1861
FROM UTAH TERRITORY
Pautah County Line
Area was proposed to be included in CALIFORNIA 1852
39th Washington Meridian 1861
FROM UTAH TERRITORY 1862
38th Washington Meridian 1862
FROM UTAH TERRITORY 1866
37th Washington Meridian 1866
To WYOMING TERRITORY from UTAH TERRITORY 1868
Green River
To US 1848
To Mexico 1821
To US 1803
To US 1845
To TEXAS 1836
To MEXICO 1821
To US 1803
Adams-Onis Line
42nd Parallel Northern limit of proposed COLONA TERRITORY 1859
Keyapaha River
1859
102nd Meridian
NEBRASKA
1867
NEBRASKA TERRITORY 1854
Platte River
Area south of Platte River was proposed to be included in KANSAS 1857
LOUISIANA
UTAH
1896
UTAH TERRITORY 1850
State of DESERET Proposed 1867
State of DESERET Proposed 1849
33rd Washington Meridian 1859
Western boundary of extralegal Jefferson Territory 1859
From UTAH TERRITORY to NEBRASKA TERRITORY 1861
FROM NEBRASKA TERRITORY 1861
103rd Meridian - East limit of proposed COLONA TERRITORY 1859
Western limit of proposed COLONA TERRITORY 1859
To US 1821
COLORADO
1876
COLORADO TERRITORY 1861
FROM UTAH TERRITORY 1861
FROM KANSAS TERRITORY 1861
Arkansas River 1821
25th Washington Meridian
Eastern boundary of extralegal Jefferson Territory 1859
KANSAS
KANSAS TERRITORY
CALIFORNIA
1850
Von Schmidt Line 1850
San Joaquin River
36°30": Proposed southern boundary of CALIFORNIA 1850
36th Parallel: Pico Line. Proposed southern boundary of CALIFORNIA 1859
37th Parallel 1850
FROM ARIZONA TERRITORY 1867
Virgin River
Virgin River was to be southern boundary of proposed NEVADA TERRITORY 1859
Missouri Compromise Line 1850
Texas Meridian
Area was proposed to be included in NEW MEXICO 1911
FROM NEW MEXICO TERRITORY 1861
FROM TEXAS 1850
Southern boundary of extralegal JEFFERSON TERRITORY 1859
Public Land Strip to OKLAHOMA TERRITORY 1890
Cherokee Outlet to Oklahoma Territory 1893
State of COLORADO Proposed from southern CALIFORNIA 1854
Proposed COLORADO TERRITORY south of Pico Line 1859
State of DESERET Proposed 1849
MEXICAN CESSION 1848
ARIZONA
1912
ARIZONA TERRITORY 1863
FROM NEW MEXICO TERRITORY
109th Meridian Proposed eastern boundary of ARIZONA TERRITORY 1859 per New Mexico Territorial Legislature
DESERET LINE 1849
NEW MEXICO
1912
NEW MEXICO TERRITORY 1850
FROM TEXAS
Rio Grande 1836
36°30" 1850
Canadian River
To US 1845
To TEXAS 1836
To MEXICO 1821
Louisiana Purchase 1803
100th Meridian
N Fk Red River
GREER CO FROM TEXAS 1896
Prairie Dog Town Fork
Bell Line 1850
103rd Meridian 1850
Colorado River
34th Parallel 1861
CONFEDERATE ARIZONA TERRITORY 1861
Gila River 1848
San Diego
Weller Line 1848
MEXICO offers to sell Baja California to US 1917
33°40' Northern boundary of proposed ARIZONA TERRITORY 1859
33°11' Northern boundary of proposed ARIZONA TERRITORY 1857
32°30" 1848
PACIFIC OCEAN
34
35
37
38
39
40
41
43
47
48
49
51

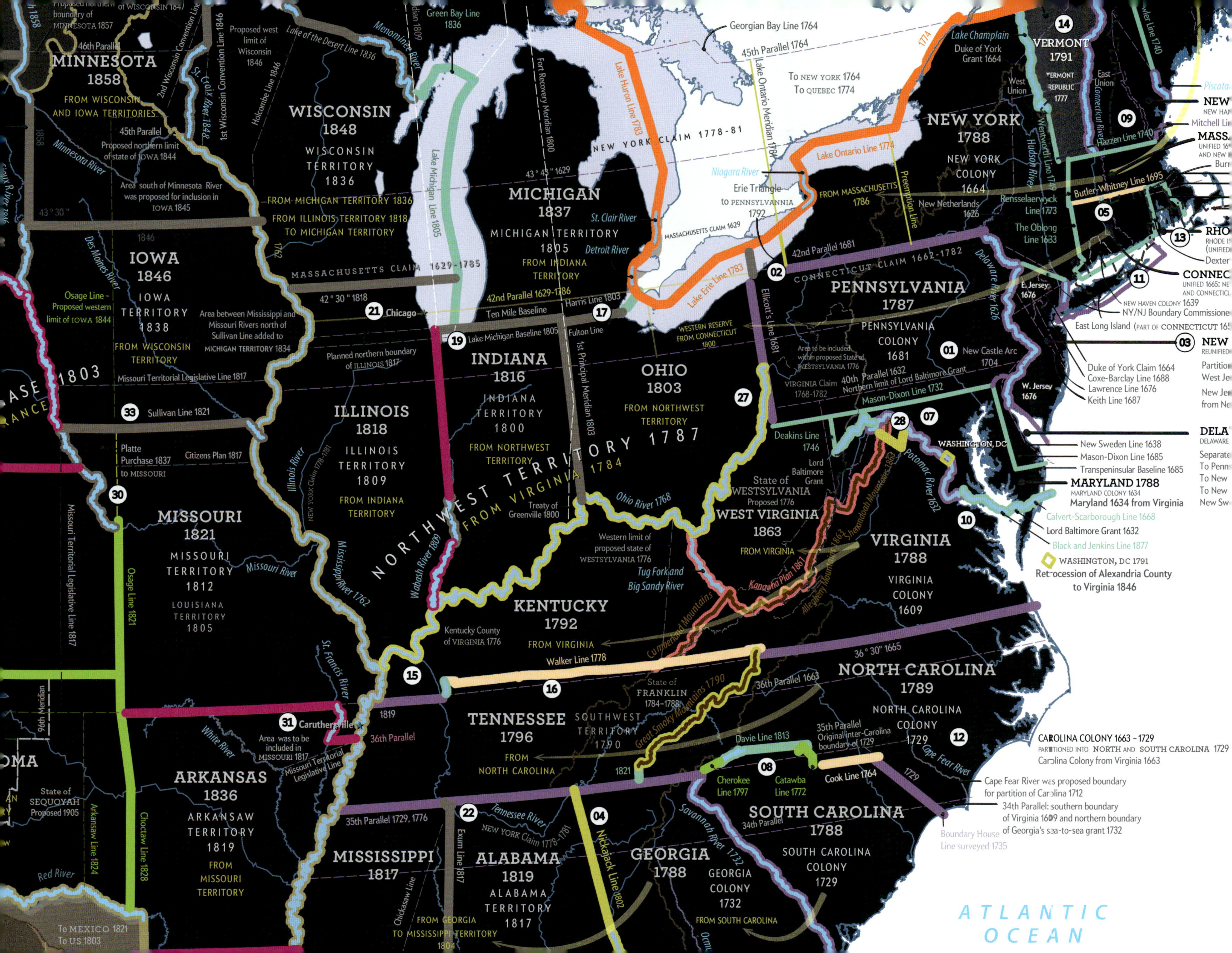
MINNESOTA
1858
FROM WISCONSIN AND IOWA TERRITORIES
46th Parallel
45th Parallel
Proposed northern limit of state of IOWA 1844
Minnesota River
Area south of Minnesota River was proposed for inclusion in IOWA 1845
43°30"
Proposed west limit of Wisconsin 1846
Lake of the Desert Line 1836
Menominee River
Green Bay Line 1836
2nd Wisconsin Convention Line
1st Wisconsin Convention Line 1846
St. Croix River 1848
Holcombe Line 1846
WISCONSIN
1848
WISCONSIN TERRITORY 1836
FROM MICHIGAN TERRITORY 1836
FROM ILLINOIS TERRITORY 1818 TO MICHIGAN TERRITORY
MASSACHUSETTS CLAIM 1629-1785
42°30" 1818
Chicago
IOWA
1846
IOWA TERRITORY 1838
FROM WISCONSIN TERRITORY
Des Moines River
Osage Line - Proposed western limit of IOWA 1844
Area between Mississippi and Missouri Rivers north of Sullivan Line added to MICHIGAN TERRITORY 1834
Missouri Territorial Legislative Line 1817
Sullivan Line 1821
Planned northern boundary of ILLINOIS 1817
Platte Purchase 1837
Citizens Plan 1817
to MISSOURI
Missouri Territorial Legislative Line 1817
Osage Line 1821
MISSOURI
1821
MISSOURI TERRITORY 1812
LOUISIANA TERRITORY 1805
Missouri River
Lake Michigan Line 1805
Fort Recovery Meridian 1800
43°43" 1629
MICHIGAN
1837
MICHIGAN TERRITORY 1805
FROM INDIANA TERRITORY
St. Clair River
Detroit River
42nd Parallel 1629-1786
Ten Mile Baseline
Harris Line 1803
Lake Michigan Baseline 1805
Fulton Line
1st Principal Meridian 1803
INDIANA
1816
INDIANA TERRITORY 1800
FROM NORTHWEST TERRITORY
ILLINOIS
1818
ILLINOIS TERRITORY 1809
FROM INDIANA TERRITORY
Illinois River
NEW YORK Claim 1778-1781
NORTHWEST TERRITORY 1787
FROM VIRGINIA 1784
Treaty of Greenville 1800
Wabash River 1809
Mississippi River 1762
OHIO
1803
FROM NORTHWEST TERRITORY
WESTERN RESERVE FROM CONNECTICUT 1800
Ohio River 1768
Western limit of proposed state of WESTSYLVANIA 1776
Tug Fork and Big Sandy River
KENTUCKY
1792
FROM VIRGINIA
Kentucky County of VIRGINIA 1776
Walker Line 1778
Cumberland Mountains
Georgian Bay Line 1764
45th Parallel 1764
Lake Huron Line 1783
NEW YORK CLAIM 1778-81
Lake Ontario Meridian 1781
To NEW YORK 1764
To QUEBEC 1774
1774
Lake Champlain
Duke of York Grant 1664
Niagara River
Erie Triangle to PENNSYLVANIA 1792
MASSACHUSETTS CLAIM 1629
Lake Ontario Line 1774
Lake Erie Line 1783
42nd Parallel 1681
CONNECTICUT CLAIM 1662-1782
PENNSYLVANIA
1787
PENNSYLVANIA COLONY 1681
Ellicott's Line 1681
Area to be included within proposed State of WESTSYLVANIA 1776
VIRGINIA Claim 1768-1782
40th Parallel 1632
Northern limit of Lord Baltimore Grant
Mason-Dixon Line 1732
New Castle Arc 1704
Delaware River 1626
E. Jersey 1676
W. Jersey 1676
NEW YORK
1788
NEW YORK COLONY 1664
FROM MASSACHUSETTS 1786
New Netherlands 1626
Preemption Line
Hudson River
Rensselaerwyck Line 1773
The Oblong Line 1633
Wentworth Line 1749
West Union
East Union
Connecticut River
VERMONT
1791
VERMONT REPUBLIC 1777
Hazen Line 1740
Butler-Whitney Line 1695
Deakins Line 1746
Lord Baltimore Grant
State of WESTSYLVANIA Proposed 1776
WEST VIRGINIA
1863
FROM VIRGINIA
Kanawha Plan 1861
Shenandoah Mountains 1863
Allegheny Mountains
Potomac River 1632
WASHINGTON, DC
VIRGINIA
1788
VIRGINIA COLONY 1609
36°30" 1665
36th Parallel 1663
State of FRANKLIN 1784-1788
Great Smoky Mountains 1790
NORTH CAROLINA
1789
NORTH CAROLINA COLONY 1729
35th Parallel Original inter-Carolina boundary of 1729
Cape Fear River
1729
Davie Line 1813
1821
Cherokee Line 1797
Catawba Line 1772
Cook Line 1764
TENNESSEE
1796
SOUTHWEST TERRITORY 1790
FROM NORTH CAROLINA
1819
36th Parallel
Caruthersville
Area was to be included in MISSOURI 1817
Missouri Territorial Legislative Line
St. Francis River
White River
ARKANSAS
1836
ARKANSAW TERRITORY 1819
FROM MISSOURI TERRITORY
96th Meridian
Missouri Territorial Legislative Line 1817
State of SEQUOYAH Proposed 1905
Arkansaw Line 1824
Choctaw Line 1828
Red River
To MEXICO 1821
To US 1803
35th Parallel 1729, 1776
Exum Line 1817
MISSISSIPPI
1817
Chickasaw Line
FROM GEORGIA TO MISSISSIPPI TERRITORY 1804
Tennessee River
NEW YORK Claim 1778-1781
ALABAMA
1819
ALABAMA TERRITORY 1817
Nickajack Line 1802
GEORGIA
1788
GEORGIA COLONY 1732
FROM SOUTH CAROLINA
Savannah River 1732
34th Parallel
SOUTH CAROLINA
1788
SOUTH CAROLINA COLONY 1729
Boundary House Line surveyed 1735
ATLANTIC OCEAN
New Sweden Line 1638
Mason-Dixon Line 1685
Transpeninsular Baseline 1685
MARYLAND 1788
MARYLAND COLONY 1634
Maryland 1634 from Virginia
Calvert-Scarborough Line 1668
Lord Baltimore Grant 1632
Black and Jenkins Line 1877
WASHINGTON, DC 1791
Ret-ocession of Alexandria County to Virginia 1846
CAROLINA COLONY 1663 - 1729
PARTITIONED INTO NORTH AND SOUTH CAROLINA 1729
Carolina Colony from Virginia 1663
Cape Fear River was proposed boundary for partition of Carolina 1712
34th Parallel: southern boundary of Virginia 1609 and northern boundary of Georgia's sea-to-sea grant 1732
Duke of York Claim 1664
Coxe-Barclay Line 1688
Lawrence Line 1676
Keith Line 1687
NEW HAVEN COLONY 1639
NY/NJ Boundary Commissioners
East Long Island (PART OF CONNECTICUT)
01
02
03
04
05
07
08
09
10
11
12
13
14
15
16
17
19
21
22
27
28
30
31
33

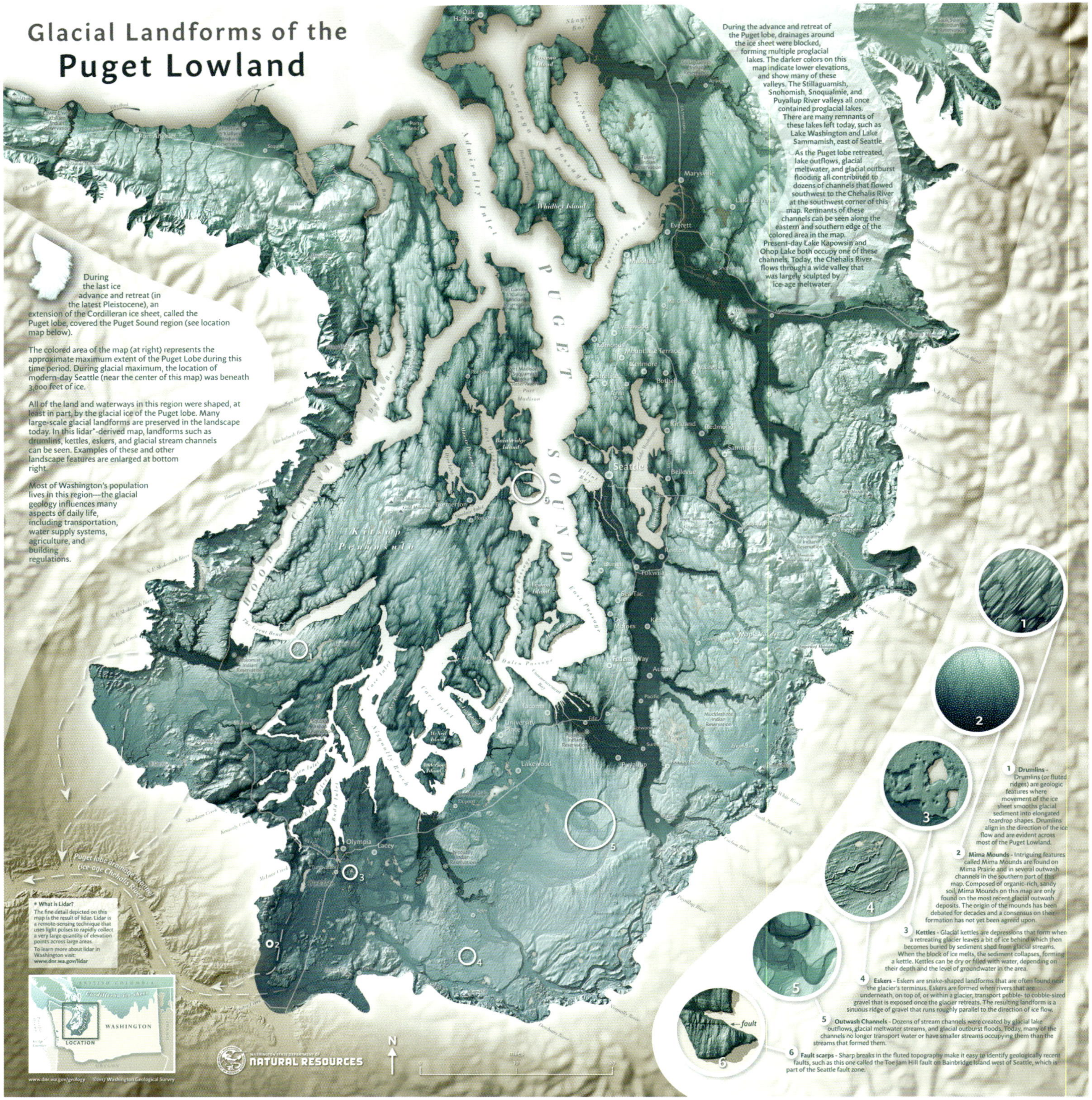
Glacial Landforms of the
Puget Lowland
During the last ice advance and retreat (in the latest Pleistocene), an extension of the Cordilleran ice sheet, called the Puget lobe, covered the Puget Sound region (see location map below).
The colored area of the map (at right) represents the approximate maximum extent of the Puget Lobe during this time period. During glacial maximum, the location of modern-day Seattle (near the center of this map) was beneath 3,000 feet of ice.
All of the land and waterways in this region were shaped, at least in part, by the glacial ice of the Puget lobe. Many large-scale glacial landforms are preserved in the landscape today. In this lidar*-derived map, landforms such as drumlins, kettles, eskers, and glacial stream channels can be seen. Examples of these and other landscape features are enlarged at bottom right.
Most of Washington's population lives in this region—the glacial geology influences many aspects of daily life, including transportation, water supply systems, agriculture, and building regulations.
During the advance and retreat of the Puget lobe, drainages around the ice sheet were blocked, forming multiple proglacial lakes. The darker colors on this map indicate lower elevations, and show many of these valleys. The Stillaguamish, Snohomish, Snoqualmie, and Puyallup River valleys all once contained proglacial lakes. There are many remnants of these lakes left today, such as Lake Washington and Lake Sammamish, east of Seattle.
As the Puget lobe retreated, lake outflows, glacial meltwater, and glacial outburst flooding all contributed to dozens of channels that flowed southwest to the Chehalis River at the southwest corner of this map. Remnants of these channels can be seen along the eastern and southern edge of the colored area in the map. Present-day Lake Kapowsin and Ohop Lake both occupy one of these channels. Today, the Chehalis River flows through a wide valley that was largely sculpted by ice-age meltwater.
PUGET SOUND
HOOD CANAL
Admiralty Inlet
Whidbey Island
Kitsap Peninsula
Marysville
Everett
Seattle
Bellevue
Kirkland
Redmond
Bremerton
Tacoma
Lakewood
Olympia
Lacey
Kent
Auburn
Puget lobe drainage channel (ice-age Chehalis River)
* What is Lidar?
The fine detail depicted on this map is the result of lidar. Lidar is a remote-sensing technique that uses light pulses to rapidly collect a very large quantity of elevation points across large areas.
To learn more about lidar in Washington visit: www.dnr.wa.gov/lidar
BRITISH COLUMBIA
WASHINGTON
LOCATION
OREGON
www.dnr.wa.gov/geology ©2017 Washington Geological Survey
NATURAL RESOURCES
N
1 Drumlins - Drumlins (or fluted ridges) are geologic features where movement of the ice sheet smooths glacial sediment into elongated teardrop shapes. Drumlins align in the direction of the ice flow and are evident across most of the Puget Lowland.
2 Mima Mounds - Intriguing features called Mima Mounds are found on Mima Prairie and in several outwash channels in the southern part of this map. Composed of organic-rich, sandy soil, Mima Mounds on this map are only found on the most recent glacial outwash deposits. The origin of the mounds has been debated for decades and a consensus on their formation has not yet been agreed upon.
3 Kettles - Glacial kettles are depressions that form when a retreating glacier leaves a bit of ice behind which then becomes buried by sediment shed from glacial streams. When the block of ice melts, the sediment collapses, forming a kettle. Kettles can be dry or filled with water, depending on their depth and the level of groundwater in the area.
4 Eskers - Eskers are snake-shaped landforms that are often found near the glacier's terminus. Eskers are formed when rivers that are underneath, on top of, or within a glacier, transport pebble- to cobble-sized gravel that is exposed once the glacier retreats. The resulting landform is a sinuous ridge of gravel that runs roughly parallel to the direction of ice flow.
5 Outwash Channels - Dozens of stream channels were created by glacial lake outflows, glacial meltwater streams, and glacial outburst floods. Today, many of the channels no longer transport water or have smaller streams occupying them than the streams that formed them.
6 Fault scarps - Sharp breaks in the fluted topography make it easy to identify geologically recent faults, such as this one called the Toe Jam Hill fault on Bainbridge Island west of Seattle, which is part of the Seattle fault zone.
fault

« Glacial Landforms of the Puget Lowland

Daniel Coe — Washington Geological Survey

During the last Ice Age, the Puget lobe of the Cordilleran ice sheet extended southward over the Puget Lowland region of Washington. To this day, the sheet's icy fingerprint can be seen on the landscape. I created this map to explain some of the fascinating landforms left behind.

I designed the focal point of the map to correspond with the extent of the ice sheet and colored the land with chilly blues to evoke the cold crevasses of an ancient glacier. Instead of using a traditional blue for the waterways of Puget Sound, I increased the contrast between land and water by representing the waterways with the milky gray color of streams fed by glaciers and tinged with glacial flour.

In many maps, terrain is the underlying base for the main subject matter, but in this map, the terrain is the primary subject, and the visual hierarchy reflects this fact. I created labels that are partially transparent—legible at close range, but less visually prominent when viewing the map as a whole. I relocated river labels beyond the glaciated boundary where they do not obscure the landforms. I also vertically exaggerated the lidar-based topography and added shadows to the shaded relief to bring out the details in the landscape.

To add interpretive value and to encourage viewers to explore the map further, I added six geomorphic examples from the landscape that are enlarged and explained. Additional text is wrapped around the periphery of the map in forms that mimic the rounded shape of the Puget lobe.

By letting the glacial theme guide the design and focusing on the region's compelling terrain, I created a map that gives viewers a clear connection to Puget Sound's icy past.

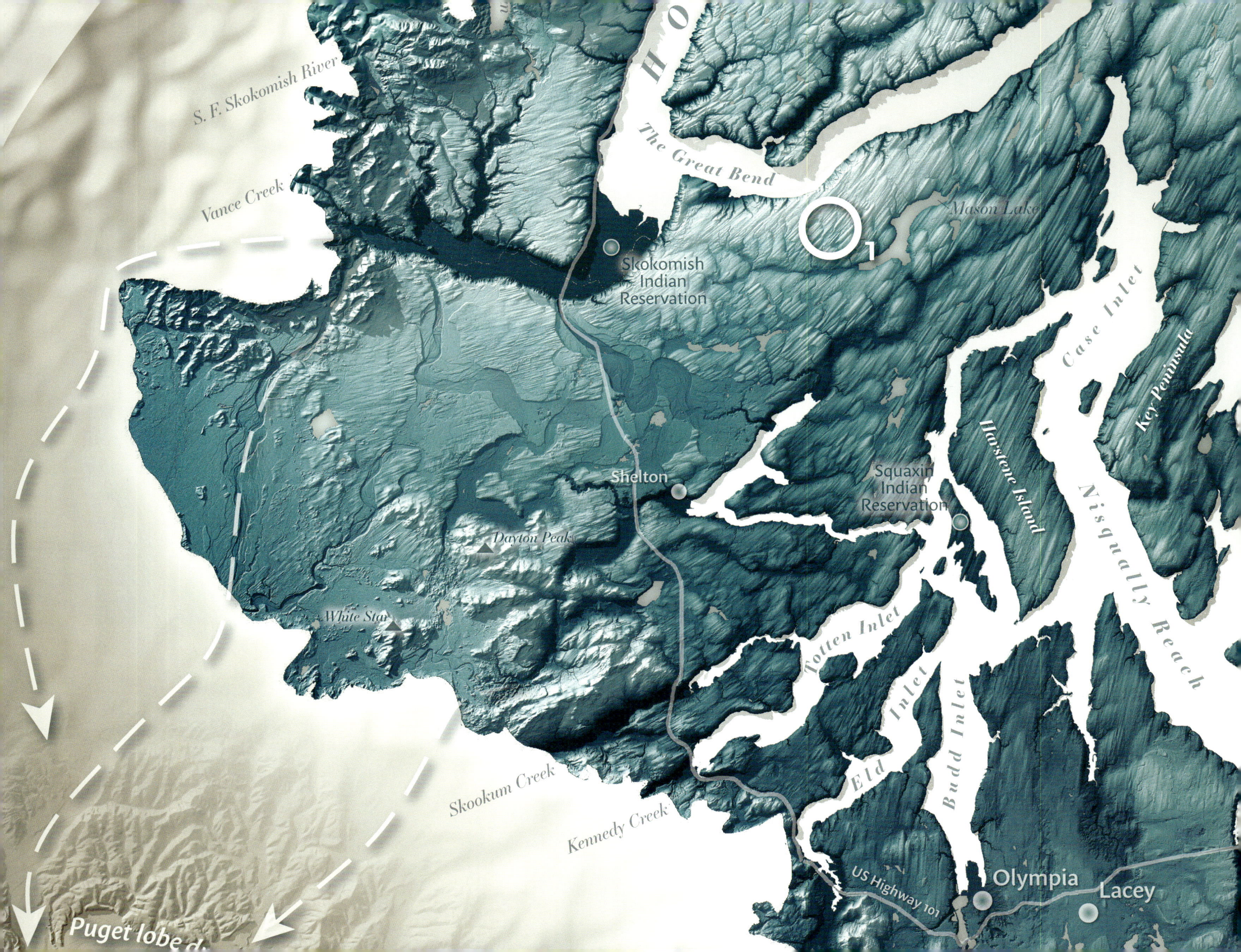

S. F. Skokomish River
The Great Bend
Vance Creek
Mason Lake
1
Skokomish
Indian
Reservation
Case Inlet
Key Peninsula
Hartstene Island
Squaxin
Indian
Reservation
Shelton
Dayton Peak
White Star
Nisqually Reach
Totten Inlet
Eld Inlet
Budd Inlet
Skookum Creek
Kennedy Creek
US Highway 101
Olympia
Lacey
Puget lobe

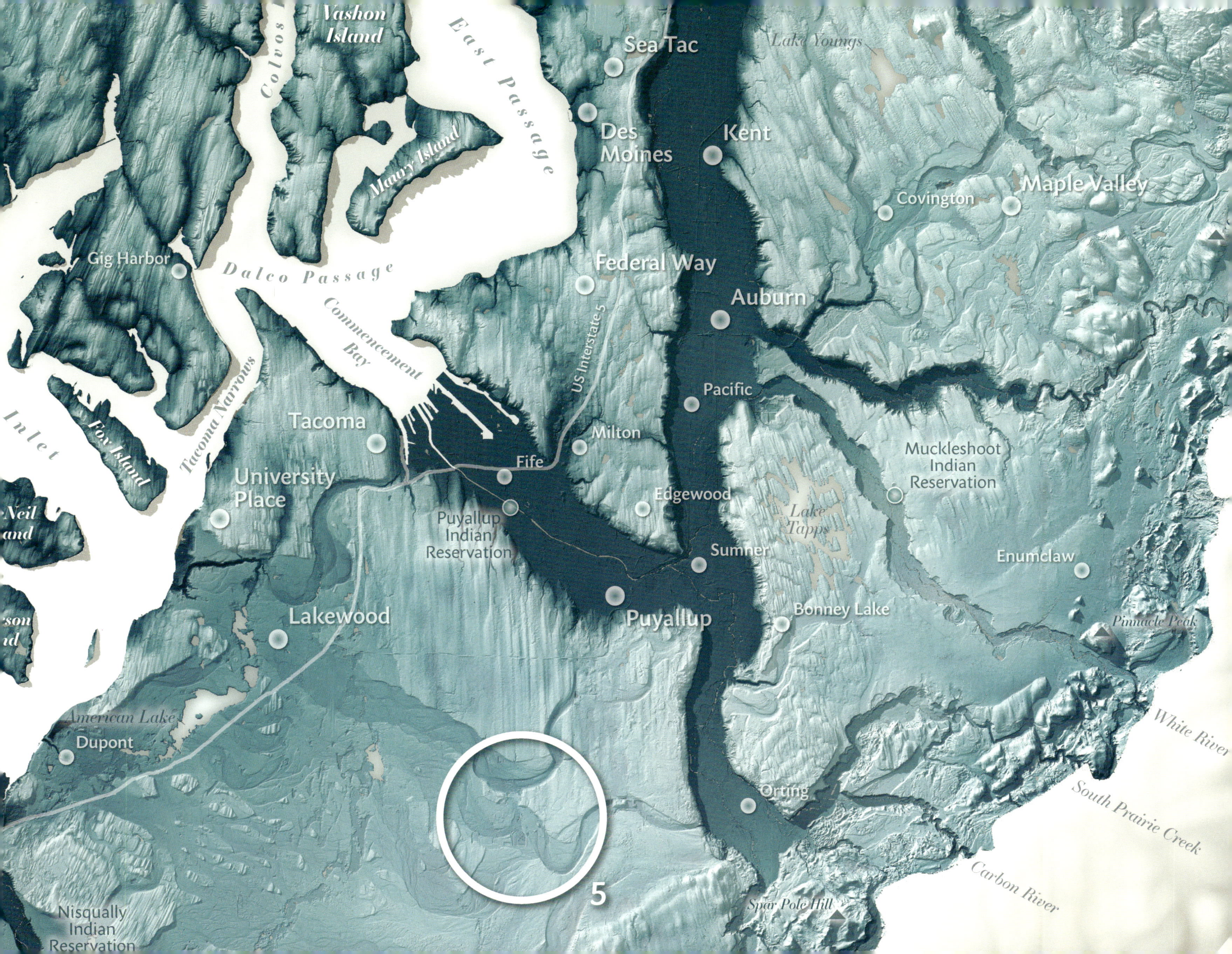
Vashon Island
Colvos
East Passage
Sea Tac
Lake Youngs
Maury Island
Des Moines
Kent
Maple Valley
Covington
Gig Harbor
Dalco Passage
Federal Way
Auburn
Commencement Bay
US Interstate 5
Inlet
Fox Island
Tacoma Narrows
Tacoma
Pacific
Milton
Fife
Muckleshoot Indian Reservation
University Place
Edgewood
Puyallup Indian Reservation
Lake Tapps
Sumner
Enumclaw
Lakewood
Puyallup
Bonney Lake
Pinnacle Peak
American Lake
Dupont
White River
Orting
South Prairie Creek
5
Carbon River
Spar Pole Hill
Nisqually Indian Reservation

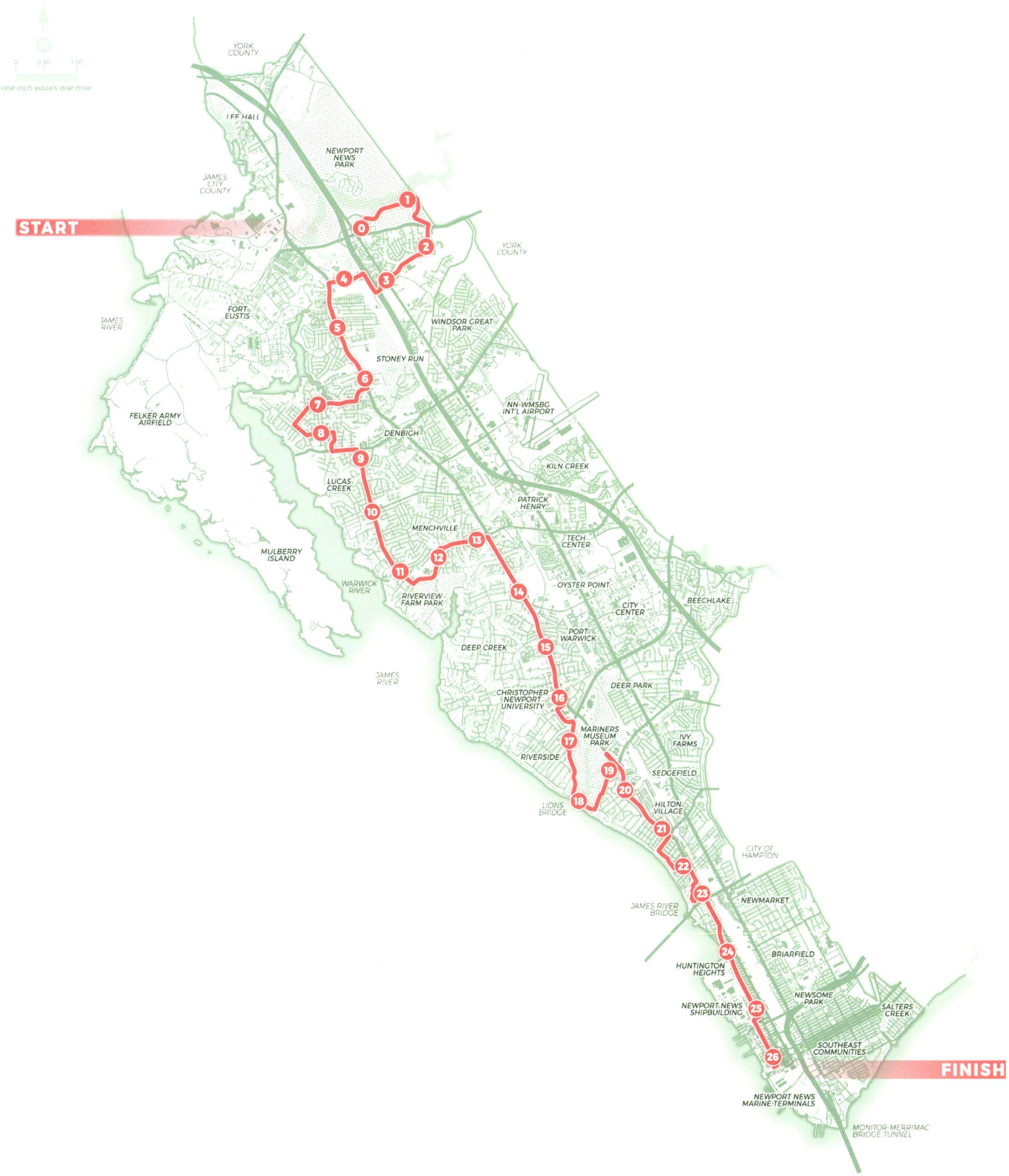

THE NEWPORT NEWS ONE CITY MARATHON

March 04, 2018 — Newport News, Virginia — www.onecitymarathon.com

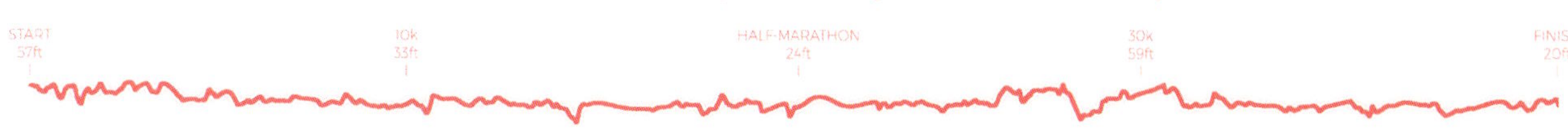

« One City Marathon

Jonah Adkins

I quit smoking in 2004 and took up running as part of a personal journey. After falling in love with a local trail, I merged my new hobby with my profession as a cartographer, creating detailed maps of a favorite running spot. In 2014, when I heard there was going to be a marathon spanning my hometown of Newport News, Virginia, I was eager to get running and mapping.

The One City Marathon is now an annual event in Newport News. Stretching nearly the entire length of the city, the marathon route takes advantage of flat terrain and long streetscapes.

As I do with most passion projects, I toiled through countless versions of what would become the One City Marathon map. In 2017, the third year of the race, my ideas finally bore fruit. Earlier versions were too detailed and complex. When I decided to keep it simple, the design finally worked.

Using a few shades of just two colors lets relevant route information pop vibrantly. Allowing the background to have a strong presence without overpowering the map required delicate balance. The race takes place along U.S. Highway 60, one of the city's main transportation routes, and cuts through many neighborhoods. The roads and buildings on the map illustrate the density of infrastructure. Newport News is surrounded by water, but it's not important to the race, so I muted the water in the color palette. A stipple pattern in the parks gives them just enough negative space to show the green space. There is also a slight hint of hillshade to complement the elevation profile below the title.

The labels on the map are specific, and few of them would take precedence on most commercial maps. My goal was that the labels should make sense to locals, introduce out-of-towners to how things are named in Newport News, and provide viewers with the same warm feeling I have about my hometown.

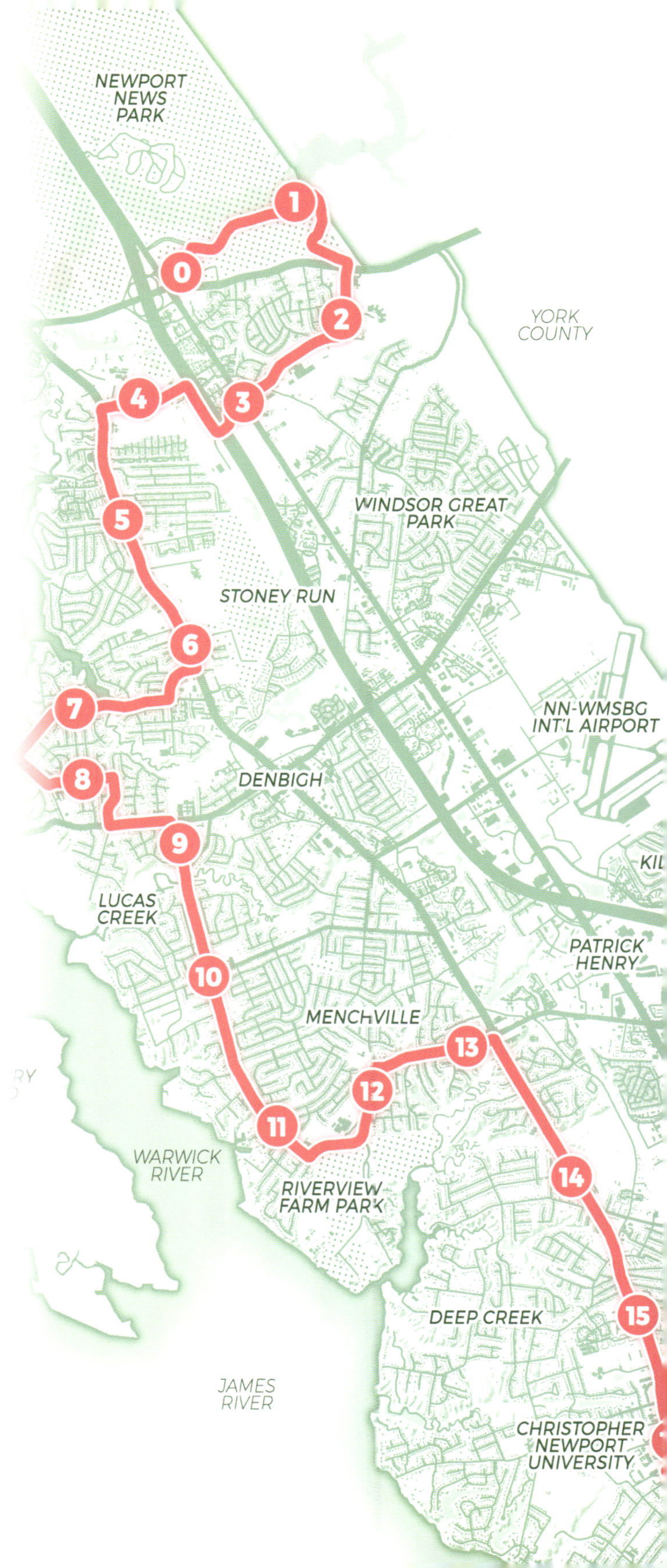

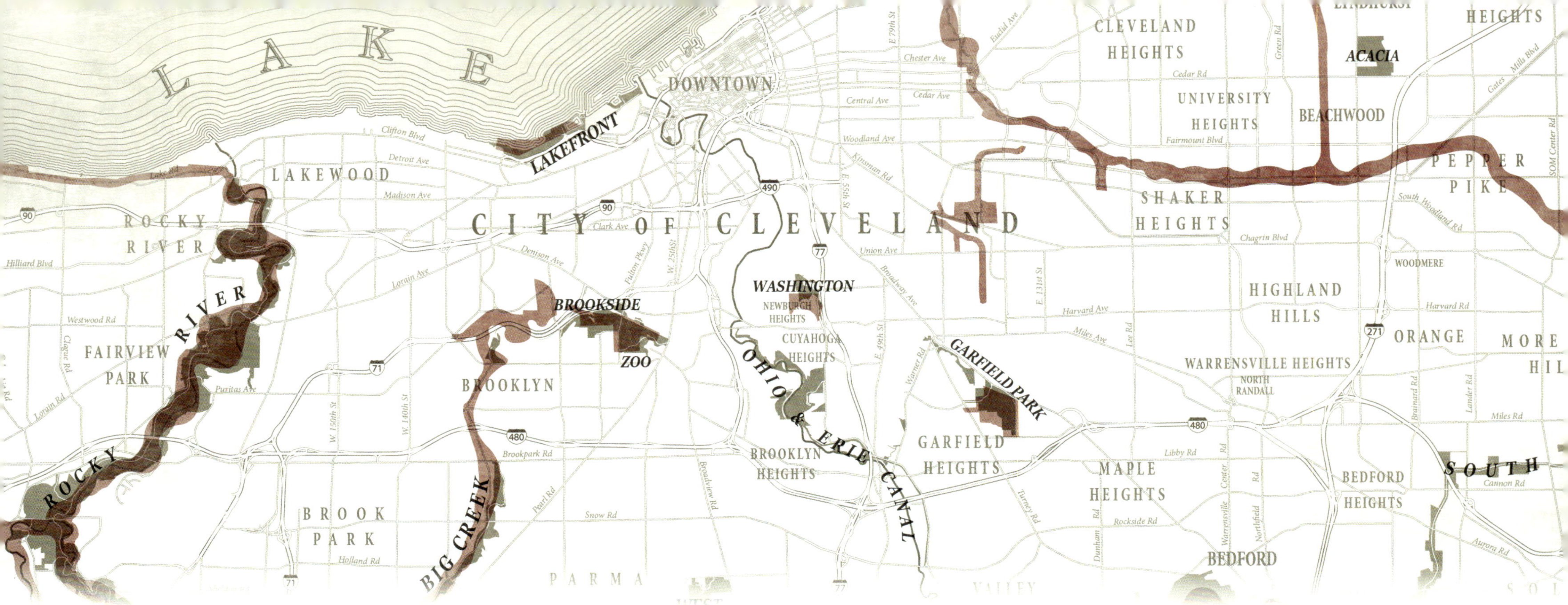

Cleveland Metroparks Centennial Map »

Brandon Garman — Cleveland Metroparks

My inspiration for the Cleveland Metroparks Centennial map came from many sources, including the original park plan from 1916 and other maps created in the early twentieth century. This map focuses on texture, color, and unique elements from the original plan that tie everything together.

The texturing of the map is critical to maintaining the worn, old map aesthetic, and the appearance of randomness helps achieve this. I used paper, scratch, and smudge textures set at different opacities and multiple blending modes to get the old, worn look. I gave each texture a hand-drawn mask to bring further control to the textural details around the map, allowing me to highlight sections, create areas of transition, and link all of the map elements together.

I have always admired the way old maps display a lot of information and tell stories without using too much color. Subtle changes in saturation and line types help to achieve the desired aesthetic and accentuate the dusky red color reserved for the most important feature on the map: the historic park plan.

From the beginning, the purpose of the map was to honor the original park plan. I used an updated version of the ornate title block with the original north arrow and scale bar and matched the typography as closely as possible. I carefully considered how to incorporate the actual original park plan into the map. In the end, I chose to overlay blown-up sections of the plan in the background, in what would have been white space. In this way, viewers who look closely are rewarded with a few of those 1916 details that bring the history of the city to life.

Cleveland Metroparks

Centennial 1917-2017

The Emerald Necklace

"THE METROPOLITAN PARKS HAVE MANY VALUES TO THE GREAT URBAN COMMUNITY... THEY BELONG TO ALL OF US, WITHOUT DISTINCTION AS TO RACE, RELIGION OR POLITICAL PARTY. RICH OR POOR, THEY BELONG TO ALL OF US, LET US EVER PROTECT AND KEEP THEM."

- William H. Stinchcomb
Executive Director 1921 to 1957

CELEBRATING THE PAST

In 1905 city parks engineer, William A. Stinchcomb called for the establishment of *"an outer system of parks and boulevards"*. He had in mind the *"... valleys of Rocky River in the west, and Chagrin River in the east."* In poetic fashion he noted that this land held *"some of the finest stretches of natural park lands to be found in the northern part of Ohio."*

The 1916 concept map visualized possible parkways and reservations and as the Cleveland Metropolitan Park System took shape, it adhered closely to the original vision. By 1929, the park system looked much like it appears on this map. Also included on the original map were several City of Cleveland parks, notably, Garfield, Edgewater, Gordon and Washington parks. By 2017, those city parks had become part of what is now known as Cleveland Metroparks, making the 1916 map a truly remarkable vision of a future that has today become a reality.

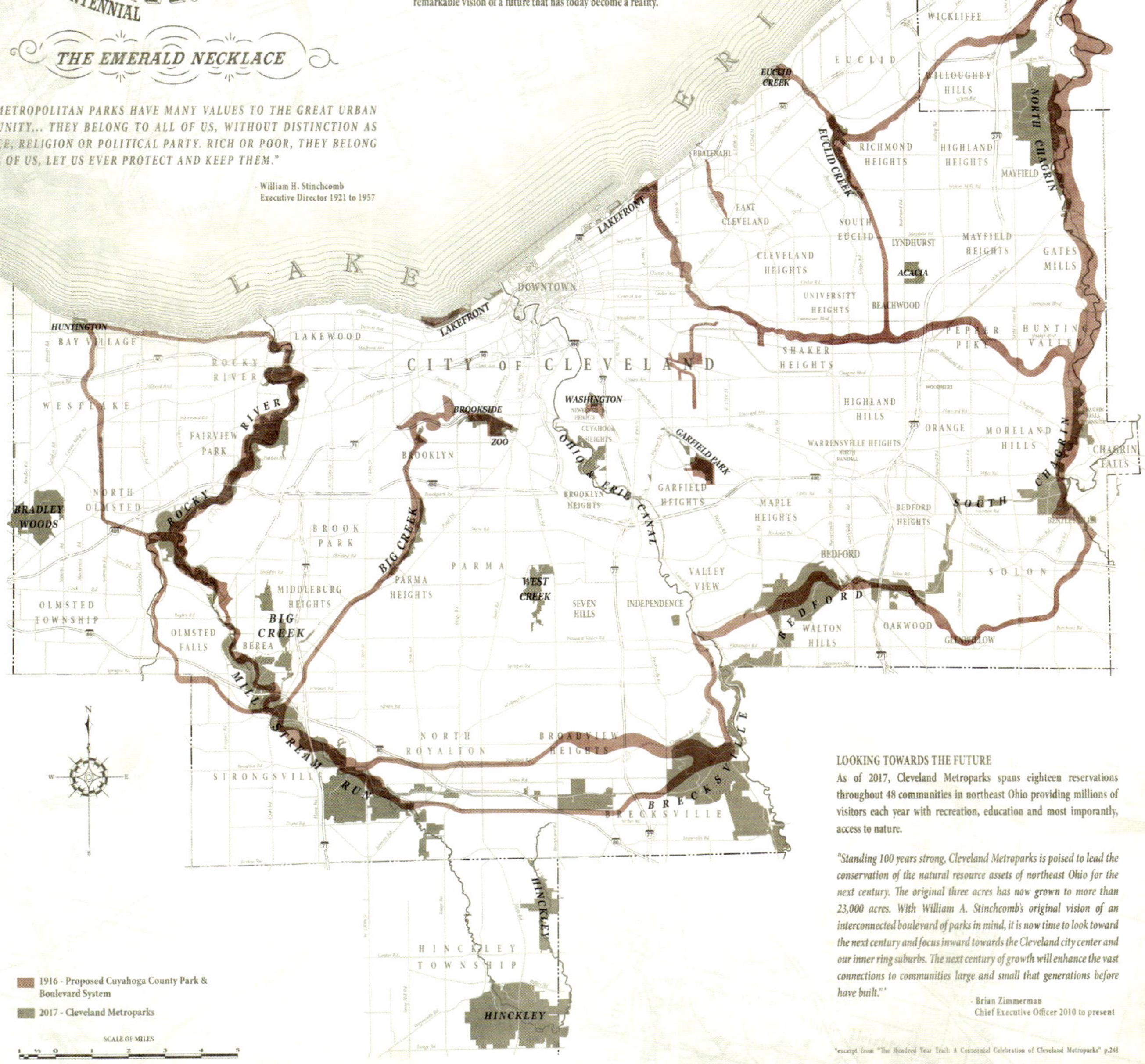

LOOKING TOWARDS THE FUTURE

As of 2017, Cleveland Metroparks spans eighteen reservations throughout 48 communities in northeast Ohio providing millions of visitors each year with recreation, education and most imporantly, access to nature.

*"Standing 100 years strong, Cleveland Metroparks is poised to lead the conservation of the natural resource assets of northeast Ohio for the next century. The original three acres has now grown to more than 23,000 acres. With William A. Stinchcomb's original vision of an interconnected boulevard of parks in mind, it is now time to look toward the next century and focus inward towards the Cleveland city center and our inner ring suburbs. The next century of growth will enhance the vast connections to communities large and small that generations before have built."**

- Brian Zimmerman
Chief Executive Officer 2010 to present

*excerpt from "The Hundred Year Trail: A Centennial Celebration of Cleveland Metroparks" p.241

King Eider »

Daniel P. Huffman

This map of the arctic life of King Eiders is among the most complicated of the 135 pieces I produced for the *Ecological Atlas of the Bering, Chukchi, and Beaufort Seas*. Its data richness posed a significant challenge: how could we overlap so many layers without everything becoming illegible? The question had no single answer, but after dozens of revisions, my colleagues at Audubon Alaska and I arrived at many partial ones.

Seasonal migration arrows have labels knocked through their bodies, revealing the underlying map and also removing the need to use yet one more color in a map already full of them. Subtle inner glows help define the boundaries of the filled polygons as they overlap and become striped. The large polygon defining the King Eider's range is left unfilled, but is gently feathered to hint at which area is inside and which outside. I exaggerated critical seabed features, leaving them easier to discern through the semitransparent color overlays.

While many small tricks like these are the foundation of the map's success, chief among them is the plethora of labels. As colors and shapes stack up, each is neatly labeled to assist a reader who may otherwise be overwhelmed by the many layers. There is almost no need to look at the legend: a label is the ultimate clarifier.

In our striving for clarity, my colleagues and I also designed for readers with color vision impairments; the fact that their needs are often neglected is, to my mind, a moral failing of the cartographic profession. I am proud that every palette in the atlas was selected to be legible to readers with protanopia, deuteranopia, or tritanopia. Meanwhile, the thorough labeling also helps reduce the need to distinguish colors in the first place.

The *Ecological Atlas* maps are some of the most challenging that I've ever worked on, requiring me to employ almost every cartographic trick that I know. But the experience pushed me to become better at my craft, and has been a deeply rewarding one.

Learn more at somethingaboutmaps.com

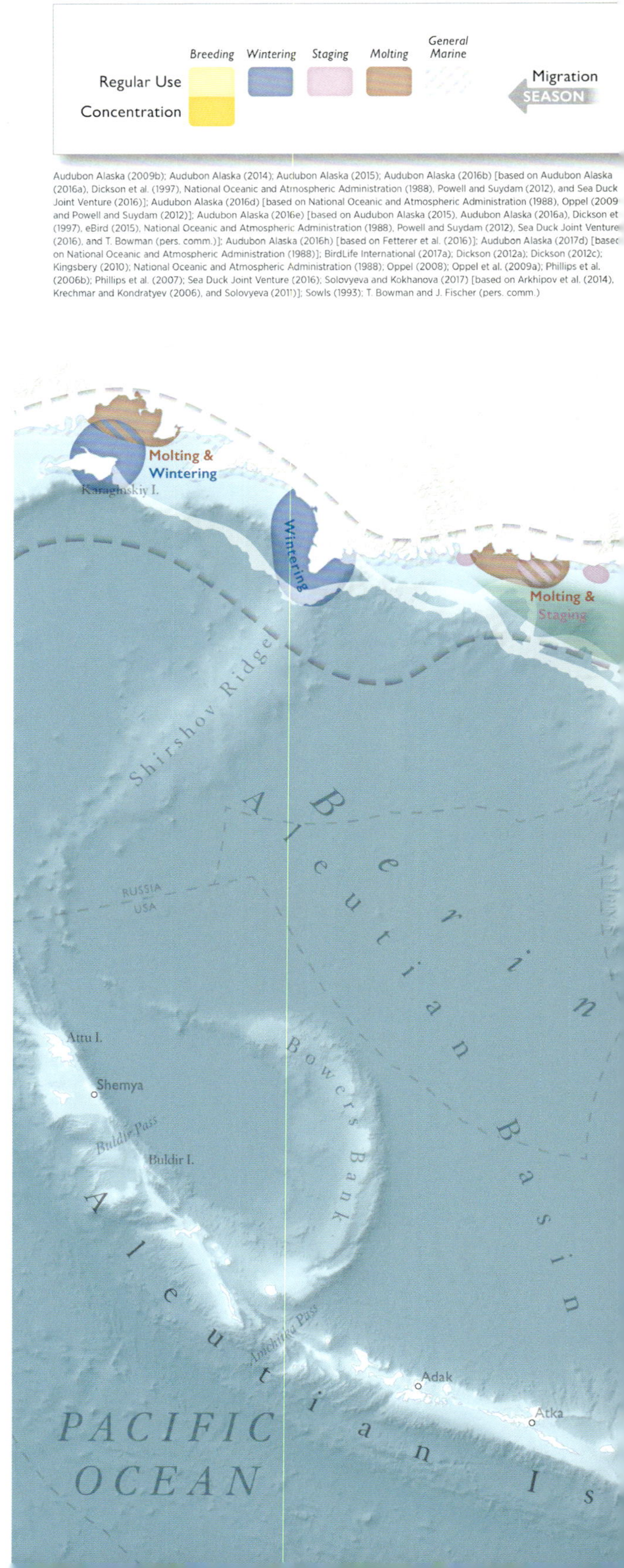

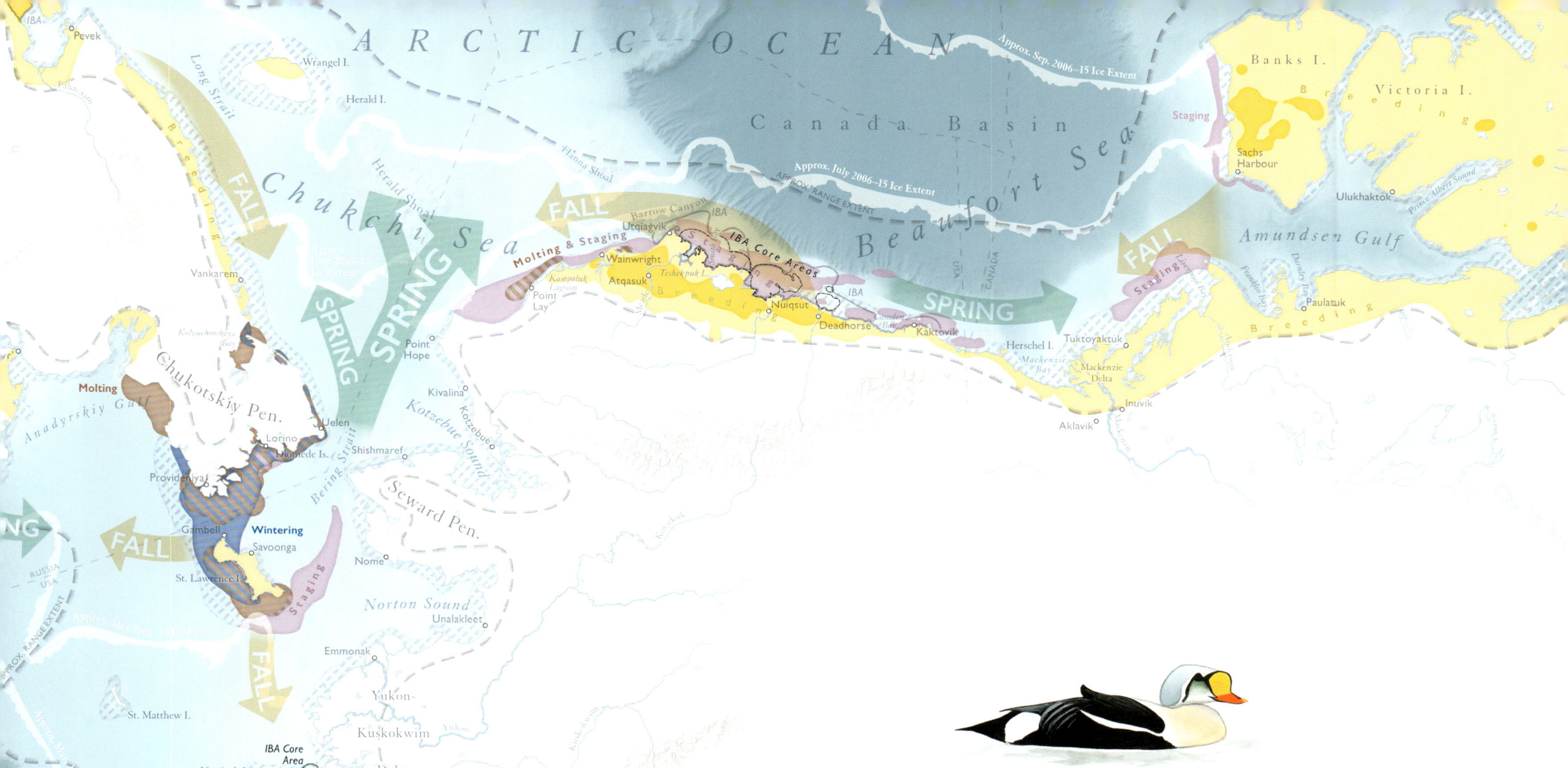

© David Allen Sibley

King Eider (*Somateria spectabilis*)

King Eiders are a conspicuous species of northern breeding sea duck found throughout the Bering, Chukchi, and Beaufort Seas. In the spring, King Eiders leave their wintering grounds in the Bering Sea in flocks of thousands, migrating north along the coast to breeding grounds on the North Slope of Alaska; the northeastern coast of Chukotka, Russia; and portions of the Canadian Arctic Archipelago. Females return to the place of their birth to nest near the shore in early May, while their nesting sites are often still frozen. They will feed in the available coastal waters as the warming weather thaws the ice. King Eiders then breed, staying close to the nest until their hatchlings fledge. The eiders stage in nearby protected, coastal waters before migrating to molting areas. Unable to fly as they molt their breeding plumage, they raft in large numbers in open water. Once their winter plumage grows in, some King Eiders will fly south to wintering areas, while others will stay in molting areas to feed as winter approaches. Many of the remaining eiders will then move south with the advancing ice edge, dispersing throughout the southern portion of their range. The King Eiders that remain concentrate in areas of open water, such as leads and polynyas, to forage throughout the winter.

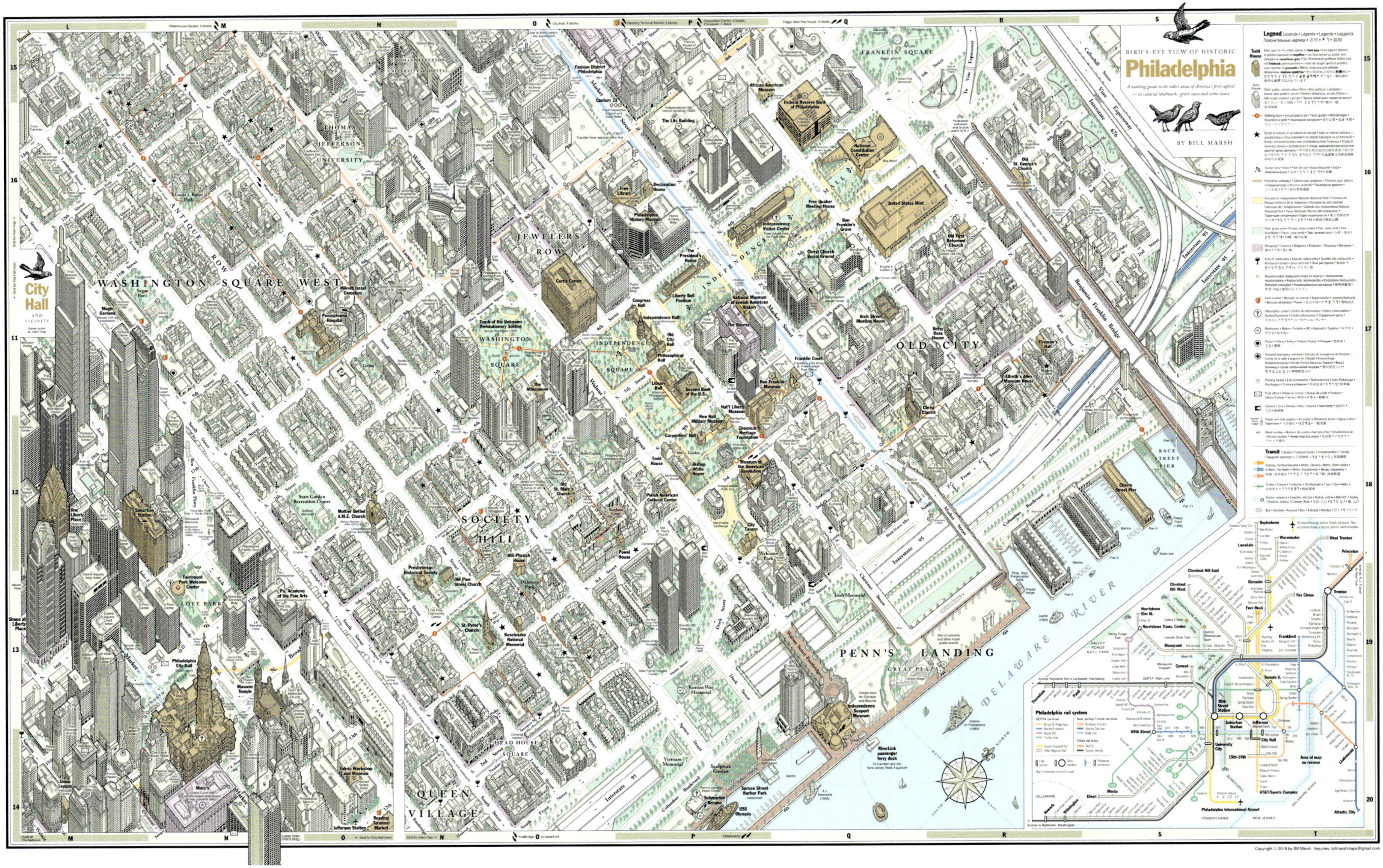
BIRD'S-EYE VIEW OF HISTORIC
Philadelphia
A walking guide to the oldest areas of America's first capital — its colonial landmarks, green oases and scenic lanes.
BY BILL MARSH
City Hall
WASHINGTON SQUARE WEST
ANTIQUE ROW
JEWELERS ROW
THOMAS JEFFERSON UNIVERSITY
WASHINGTON SQUARE
INDEPENDENCE SQUARE
INDEPENDENCE
OLD CITY
FRANKLIN SQUARE
SOCIETY HILL
HEAD HOUSE SQUARE
QUEEN VILLAGE
PENN'S LANDING
DELAWARE RIVER
LOVE PARK
Philadelphia City Hall
Masonic Temple
Reading Terminal Market
Pa. Academy of the Fine Arts
Independence Hall
Liberty Bell Pavilion
National Constitution Center
United States Mint
Federal Reserve Bank of Philadelphia
Christ Church
Elfreth's Alley
Betsy Ross House
Benjamin Franklin Bridge
Interstate 95
Interstate 676
Vine Street Expressway
Philadelphia rail system
Philadelphia International Airport
PENNSYLVANIA
NEW JERSEY
DELAWARE
Atlantic City
Legend

Much of what makes Philadelphia so interesting is hidden in small passageways and courtyards that standard maps ignore.

« Bird's-Eye View of Historic Philadelphia

Bill Marsh

Constantine Anderson created what may be the twentieth century's greatest graphic: a bird's-eye view (or, as he called it, an axonometric map) of midtown Manhattan. His stunning work meticulously depicts every building, along with street furniture, statues, and stylized trees, inked by hand with machine-like precision. He exaggerated street widths so that fewer buildings would be obscured; edited out architectural detail without sacrificing realism; tapered tall structures outward for an illusion of perspective that registers only subconsciously, a feat of draftsmanship that would drive most artists mad.

My cousin Rebecca gave me a copy of Anderson's map on my first solo visit to New York City. Years later, I produced my own bird's-eye view of Philadelphia, a map that owes most of its stylistic choices to Anderson. I struggled to get line weights and colors right, and it took a good long while just to draw, and draw, and redraw. The Philadelphia map's scale is a bit more generous, allowing me to account for nearly every window and floor. If it's a 13-story building on the ground, it's 13 stories on the map.

I started with each building's footprint, taken from city planning maps; extended two sides vertically to create the visible facades; added windows and roof detail, using scads of reference photos. Most buildings aren't simple cubes, and the more varied their form, the more work it takes to get proportions right. Many, including our florid City Hall, required several drafts.

In my version of Philadelphia, floor heights come in small, medium, and large. This standardization can cause problems because inaccurate heights, multiplied by 40 or 50 stories, produce a building out of whack with its neighbors. In those cases, it was back to the digital drawing board with customized measurements.

Much of what makes Philadelphia so interesting is hidden in small passageways and courtyards that standard maps ignore. This map draws attention to them. I peppered streets with symbols highlighting scenic areas and threaded red walking tours along the city's picturesque lanes. I even included my own cats; the living are grouped at our home, and the deceased sit, ghosted, near the animal shelter where I found them.

I hope to experiment in the future with software that permits more realism but stops short of the photographic. Still, I appreciate the old-school visual style of this map and its kinship with engravings like the 1739 Turgot map of Paris. Most of all, I value the inspiration of Constantine Anderson's masterful work.

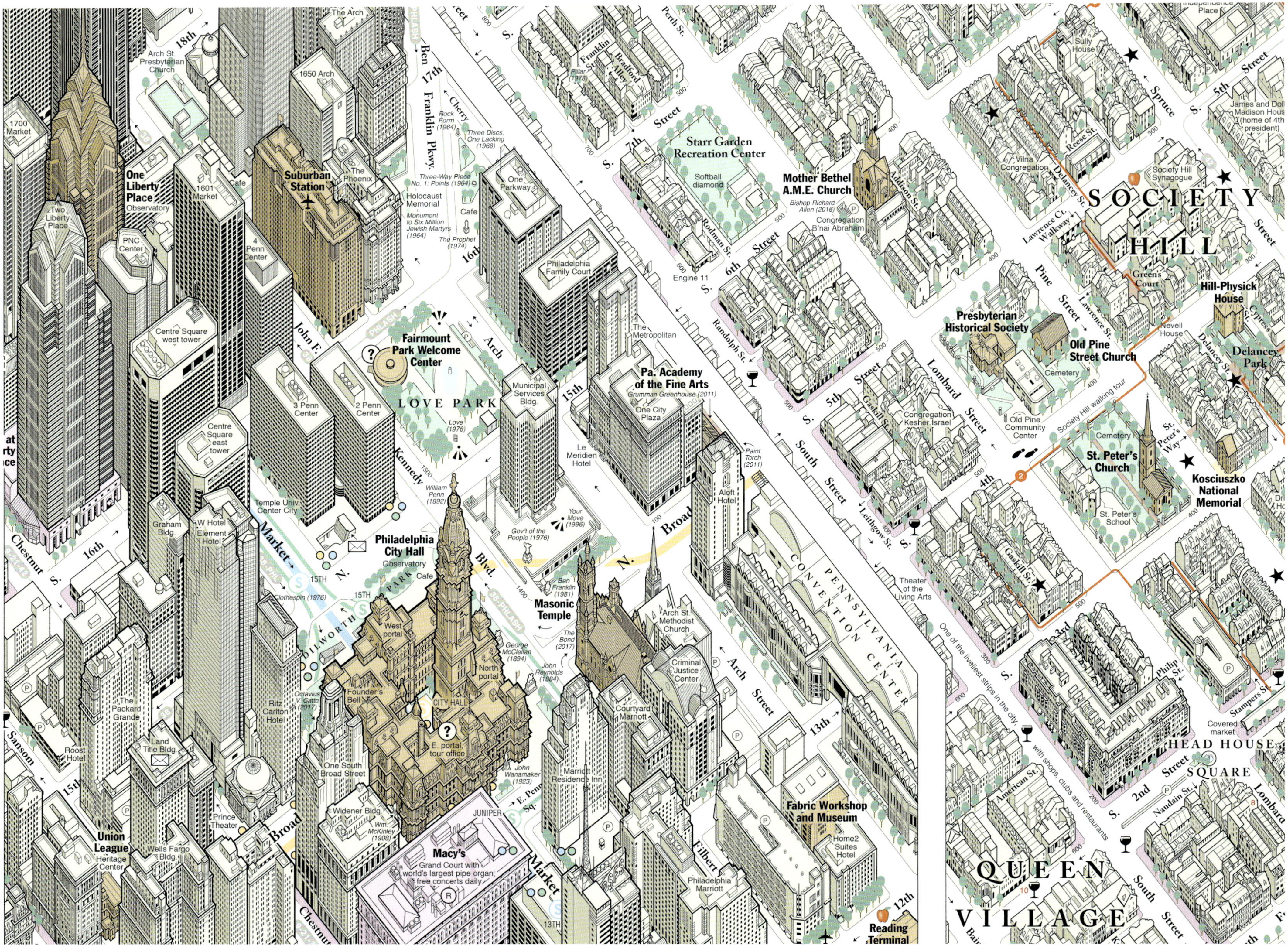

The Arch
1700 Market
Arch St. Presbyterian Church
1650 Arch
One Liberty Place
Observatory
Two Liberty Place
Suburban Station
The Phoenix
1601 Market
PNC Center
4 Penn Center
Ben Franklin Pkwy.
Cherry
Rock Form (1964)
Three Discs, One Lacking (1968)
Three-Way Piece No. 1: Points (1964)
Holocaust Memorial
Monument to Six Million Jewish Martyrs (1964)
Cafe
The Prophet (1974)
One Parkway
Philadelphia Family Court
Centre Square west tower
Fairmount Park Welcome Center
LOVE PARK
Love (1976)
3 Penn Center
2 Penn Center
Centre Square east tower
John F. Kennedy Blvd.
Arch
Municipal Services Bldg.
The Metropolitan
Pa. Academy of the Fine Arts
Grumman Greenhouse (2011)
One City Plaza
Le Meridien Hotel
William Penn (1892)
Temple Univ. Center City
Graham Bldg.
W Hotel
Element Hotel
Market
Philadelphia City Hall
Observatory
Cafe
Your Move (1996)
Gov't of the People (1976)
Ben Franklin (1981)
Masonic Temple
Paint Torch (2011)
Aloft Hotel
Broad
Chestnut
16th
15th
Clothespin (1976)
DILWORTH PARK
West portal
North portal
George McClellan (1894)
John Reynolds (1884)
The Bond (2017)
Arch St. Methodist Church
Criminal Justice Center
Founder's Bell
CITY HALL
E. portal tour office
Octavius V. Catto (2017)
Ritz Carlton Hotel
The Packard Grande
Land Title Bldg.
Roost Hotel
Sansom
One South Broad Street
Prince Theater
Union League
Heritage Center
Wells Fargo Bldg.
Widener Bldg.
Wm. McKinley (1908)
JUNIPER
John Wanamaker (1923)
Macy's
Grand Court with world's largest pipe organ; free concerts daily
Courtyard Marriott
Marriott Residence Inn
13th
Filbert
Philadelphia Marriott
Fabric Workshop and Museum
Home2 Suites Hotel
12th
Reading Terminal
PENNSYLVANIA CONVENTION CENTER
Arch Street
Franklin Sq.
Pillar (1978)
Bradford Alley
Starr Garden Recreation Center
Softball diamond
7th Street
Rodman St.
Engine 11
6th Street
Randolph St.
5th Street
Mother Bethel A.M.E. Church
Bishop Richard Allen (2016)
Congregation B'nai Abraham
Addison St.
Congregation Kesher Israel
Lombard Street
South Street
Leithgow St.
4th
Gaskill St.
Theater of the Living Arts
One of the liveliest strips in the city with shops, clubs and restaurants
Presbyterian Historical Society
Old Pine Street Church
Cemetery
Old Pine Community Center
Society Hill walking tour
Pine Street
Lawrence Ct. Walkway
Vilna Congregation
Delancey St.
Reese St.
Sully House
Spruce Street
5th Street
James and Dolly Madison House (home of 4th president)
Society Hill Synagogue
SOCIETY HILL
Greens Court
Hill-Physick House
Nevell House
Delancey Park
Cypress St.
St. Peter's Church
St. Peter's Way
Cemetery
St. Peter's School
Kosciuszko National Memorial
Independence Place
3rd
Philip St.
American St.
2nd Street
Nadaun St.
Covered market
HEAD HOUSE SQUARE
Stampers St.
Lombard
QUEEN VILLAGE
South Street

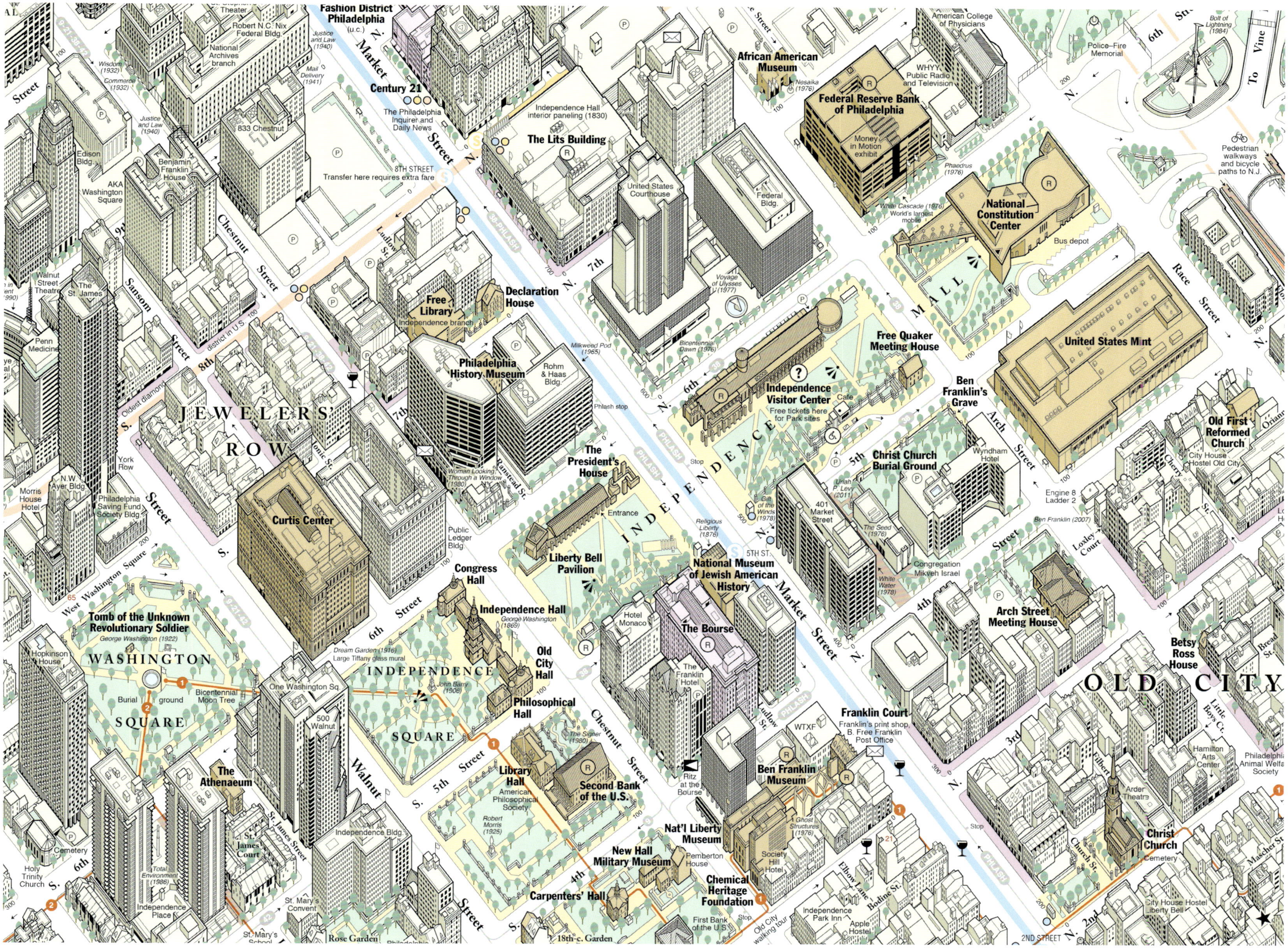

Fashion District Philadelphia
Market Street
Century 21
The Philadelphia Inquirer and Daily News
Independence Hall interior paneling (1830)
The Lits Building
African American Museum
Federal Reserve Bank of Philadelphia
Money in Motion exhibit
American College of Physicians
WHYY Public Radio and Television
Police-Fire Memorial
Bolt of Lightning (1984)
Vine
Pedestrian walkways and bicycle paths to N.J.
Robert N.C. Nix Federal Bldg.
National Archives branch
833 Chestnut
8TH STREET
Transfer here requires extra fare
Benjamin Franklin House
Edison Bldg.
AKA Washington Square
United States Courthouse
Federal Bldg.
National Constitution Center
White Cascade (1976) World's largest mobile
Bus depot
Phaedrus (1976)
MALL
Race Street
Chestnut Street
Walnut Street Theatre
The St. James
Penn Medicine
Sansom Street
Free Library Independence branch
Declaration House
Voyage of Ulysses (1977)
Bicentennial Dawn (1976)
Free Quaker Meeting House
United States Mint
Philadelphia History Museum
Rohm & Haas Bldg.
Milkweed Pod (1965)
Philash stop
Independence Visitor Center
Free tickets here for Park sites
Cafe
Ben Franklin's Grave
Arch Street
JEWELERS' ROW
Oldest diamond district in U.S.
The President's House
Christ Church Burial Ground
Wyndham Hotel
Old First Reformed Church
City House Hostel Old City
Engine 8 Ladder 2
Women Looking Through a Window (1980)
INDEPENDENCE
Entrance
401 Market Street
Unah P. Levy (2011)
The Seed (1976)
Gift of the Winds (1978)
York Row
N.W. Ayer Bldg.
Philadelphia Saving Fund Society Bldg.
Morris House Hotel
Curtis Center
Public Ledger Bldg.
Liberty Bell Pavilion
Religious Liberty (1876)
National Museum of American Jewish History
5TH ST.
Congregation Mikveh Israel
White Water (1978)
Ben Franklin (2007)
Loxley Court
West Washington Square
Congress Hall
Independence Hall
George Washington (1869)
Hotel Monaco
The Bourse
Arch Street Meeting House
Betsy Ross House
Tomb of the Unknown Revolutionary Soldier
George Washington (1922)
WASHINGTON SQUARE
Burial ground
Bicentennial Moon Tree
Hopkinson House
One Washington Sq.
500 Walnut
Dream Garden (1916) Large Tiffany glass mural
INDEPENDENCE SQUARE
John Barry (1908)
Old City Hall
Philosophical Hall
The Franklin Hotel
OLD CITY
Franklin Court
Franklin's print shop B. Free Franklin Post Office
WTXF
Ben Franklin Museum
Ritz at the Bourse
The Signer (1980)
Library Hall
American Philosophical Society
Second Bank of the U.S.
Robert Morris (1925)
The Athenaeum
St. James Court
St. James Street
Walnut Street
Independence Bldg.
Nat'l Liberty Museum
Ghost Structures (1976)
Hamilton Arts Center
Philadelphia Animal Welfare Society
Arden Theatre
Christ Church
Cemetery
New Hall Military Museum
Pemberton House
Society Hill Hotel
Elfreth's Alley
Bodine St.
Church St.
Carpenters' Hall
Chemical Heritage Foundation
First Bank of the U.S.
Old City walking tour
Independence Park Inn
Apple Hostel
2ND STREET
City House Hostel Liberty Bell
Mascher St.
Holy Trinity Church
Total Environment (1986)
Independence Place
St. Mary's Convent
St. Mary's School
Rose Garden
18th-c. Garden

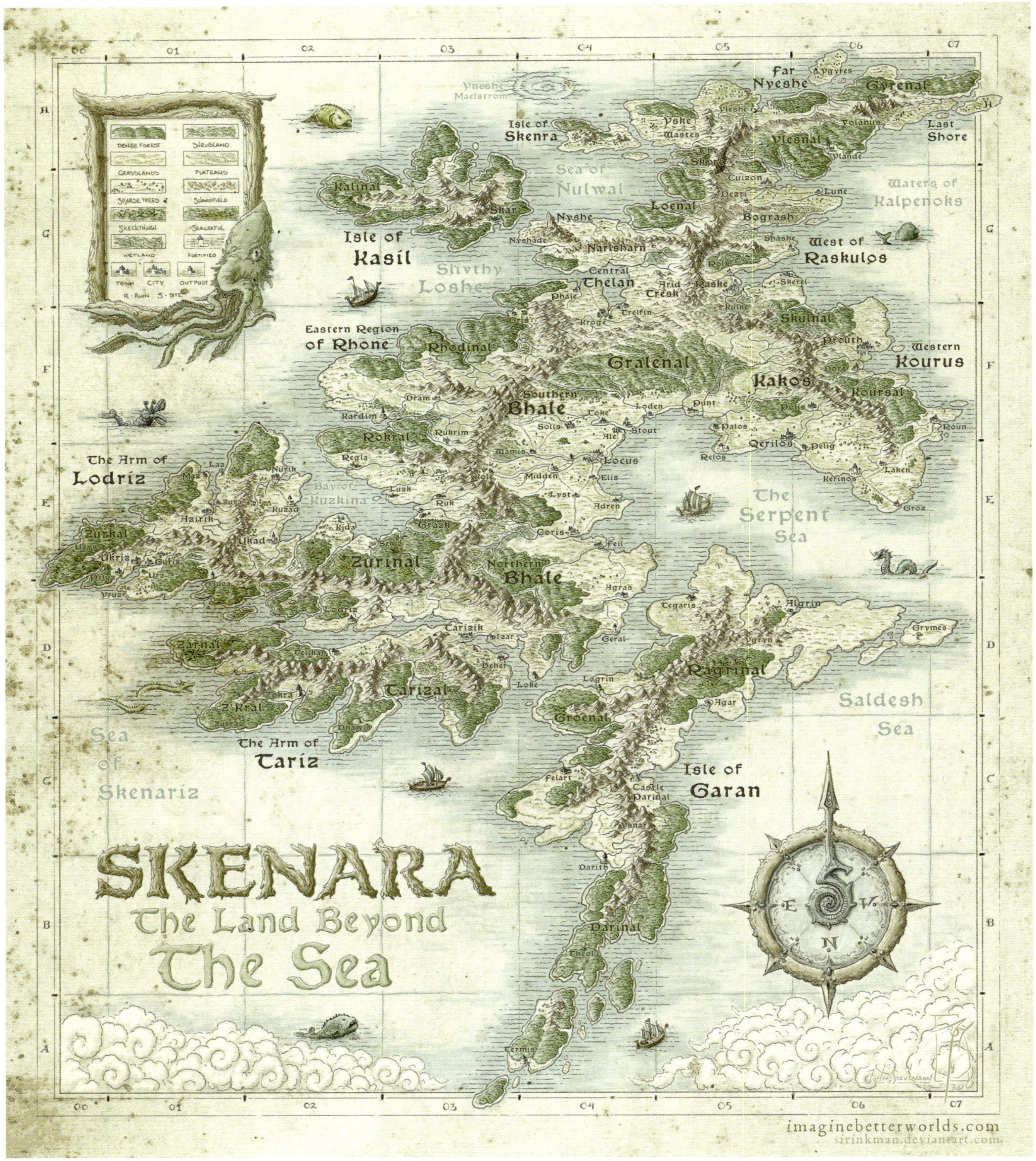
SKENARA
The Land Beyond
The Sea
Far Nyeshe
Gyrenal
Last Shore
Isle of Skenra
Sea of Nulwal
Ralinal
Skar
Isle of Kasil
Slivthy Loshe
Nyshe
Loenal
Waters of Kalpenoks
West of Raskulos
Central Thelan
Eastern Region of Rhone
Rhodinal
Gralenal
Skulnal
Western Kourus
Kakos
Southern Bhale
Rokral
The Arm of Lodriz
The Serpent Sea
Zurinal
Northern Bhale
Ragrinal
Tarizal
Saldesh Sea
Groenal
Sea of Skenariz
The Arm of Tariz
Isle of Garan
Darinal
imaginebetterworlds.com
sirinkman.deviantart.com

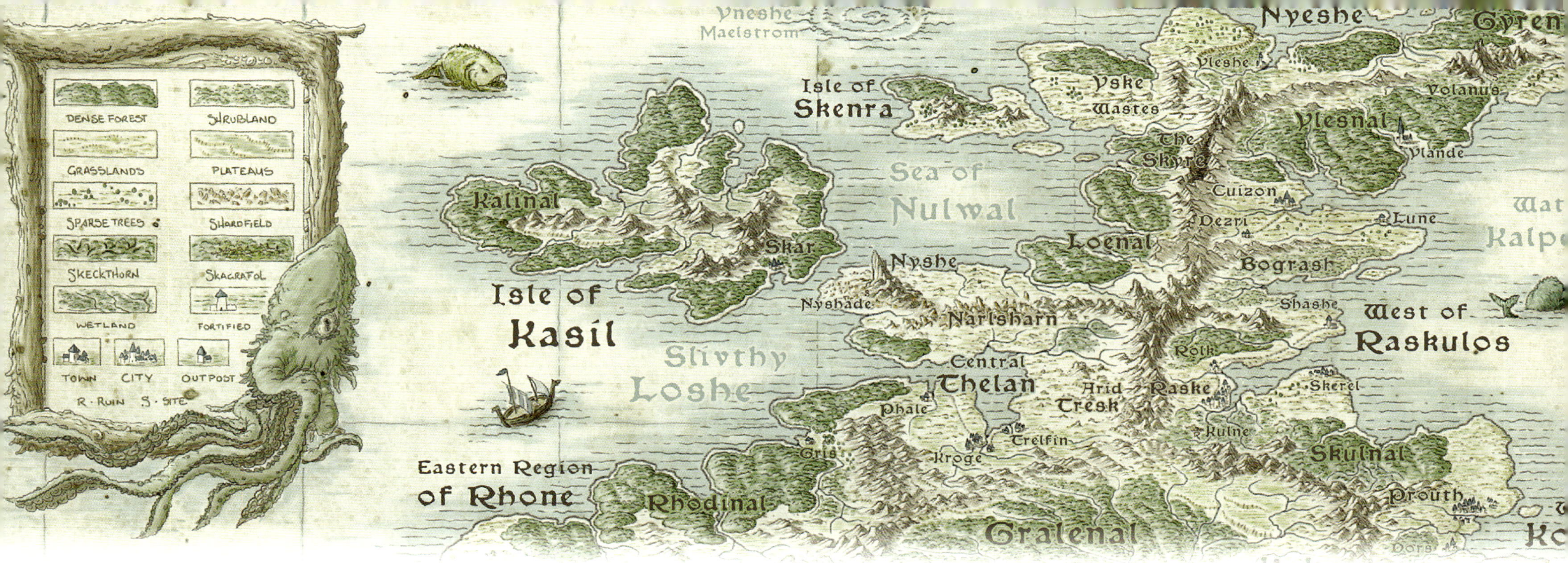

《 Skenara, The Land Beyond the Sea

John Stevenson

I am a mapmaker, illustrator, and imaginer of worlds. I frequently deal with fantastical subject matter and less frequently with real-world maps and subjects. That said, I still refer to real-world maps all the time. Looking up works by Joan Blaeu, Abraham Ortelius, and other master cartographers, I came upon an article about the depiction of sea monsters in medieval maps and was inspired to have a go at my own version.

I love the energy of early maps. I feel the sense of excitement the cartographers must have felt as they drew worlds with uncharted borders and landscapes still unknown. The motivating factor in almost all the maps I make is a fervent desire to explore the mysterious and hidden.

"Skenara, The Land Beyond the Sea," includes some of the elements of mystery that draw me to antique maps. The images push just past the edge of the frame, suggesting lands yet to explore and stories yet to tell. The sea is a theme carried throughout the map, from the cephalopod in the legend to the tentacled spiral of "due south" on the compass. The sea represents many things in this world: a means of travel, a boundary, and a never fully charted enigma.

In creating this piece, I tested how well software could mimic the look of traditional media. I crafted everything digitally, from the worn and stained paper base, to the numbers and letters on the frame, to the title lettering. I drew inspiration for my terrain rendering from the wonderful hand-drawn work of Erwin Raisz. I chose a limited color palette to enhance the overall look: green and pale blue base colors, with yellow-green highlights, soft brown shadows, and details in red-brown and darker blue.

The Skenara map was initially a test of an idea, but it has become the source of the fictional world of Nhyn, where adventurous scribes roam the lands on and beyond the map recording the world's stories.

Learn more at imaginebetterworlds.com

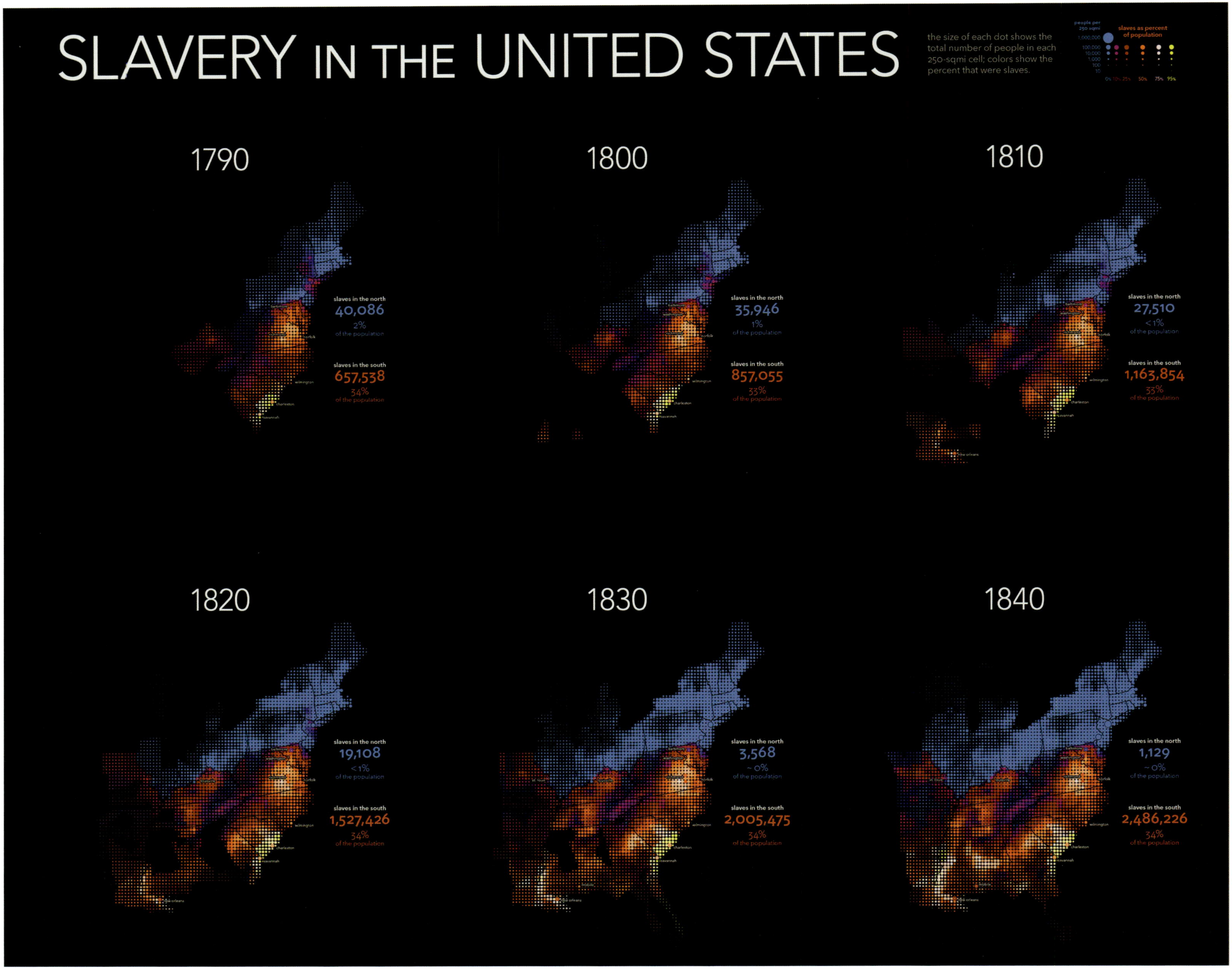
SLAVERY IN THE UNITED STATES
the size of each dot shows the total number of people in each 250-sqmi cell; colors show the percent that were slaves.
people per 250 sqmi
1,000,000
100,000
10,000
1,000
100
10
slaves as percent of population
0% 10% 25% 50% 75% 95%
1790
slaves in the north
40,086
2%
of the population
slaves in the south
657,538
34%
of the population
1800
slaves in the north
35,946
1%
of the population
slaves in the south
857,055
33%
of the population
1810
slaves in the north
27,510
<1%
of the population
slaves in the south
1,163,854
33%
of the population
1820
slaves in the north
19,108
<1%
of the population
slaves in the south
1,527,426
34%
of the population
1830
slaves in the north
3,568
~0%
of the population
slaves in the south
2,005,475
34%
of the population
1840
slaves in the north
1,129
~0%
of the population
slaves in the south
2,486,226
34%
of the population

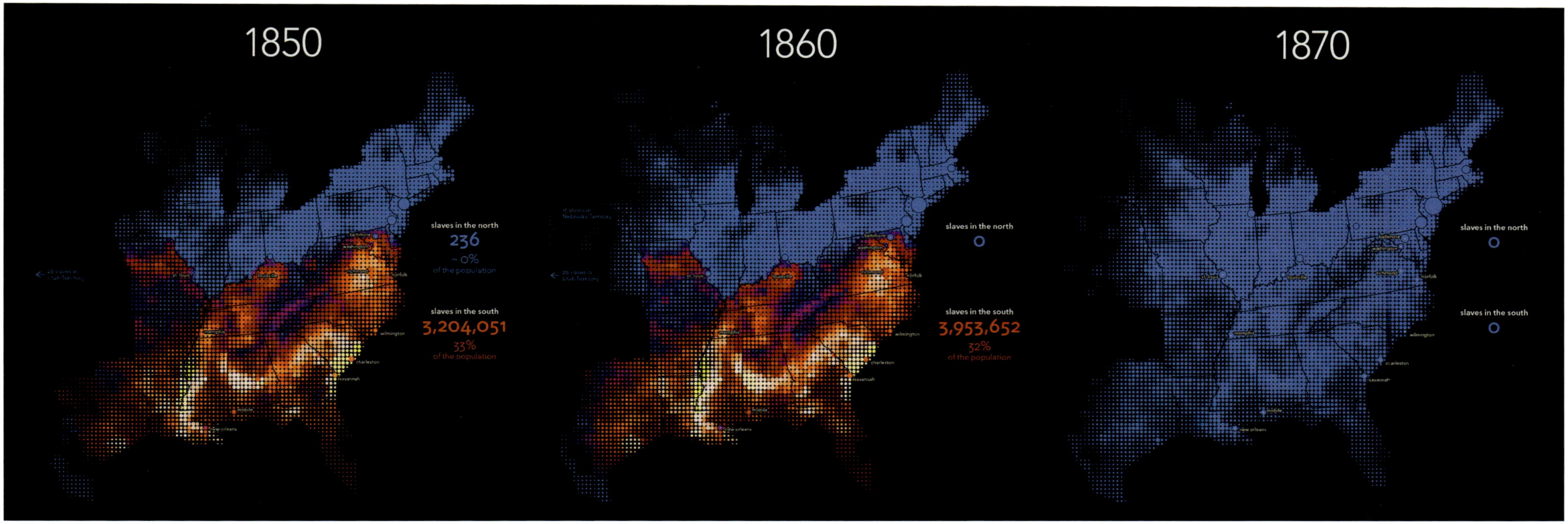

« Slavery in the United States

Bill Rankin — Yale University

Slavery is one of the most mapped topics in American history. It was the subject of some of the earliest statistical maps created in the United States, with an 1861 map by Edwin Hergesheimer famously guiding Abraham Lincoln's strategy during the Civil War, and it has remained a staple of historical atlases and textbooks ever since.

Most maps of slavery, then and now, are choropleth maps of the usual flavor: jigsaw puzzles of solid-colored shapes, with every county shaded to show enslaved people as a percentage of total population. But these maps make it impossible to see some of the most important patterns, since cities are completely invisible, comparisons with the North are ignored, changing county boundaries make temporal analysis difficult, and relative population (enslaved as a percentage of the total) takes precedence over absolute population (number enslaved).

These maps use what I call a "bubble grid" to show the geography of slavery in a new way. Instead of shading individual counties, I've aggregated population into 250-square-mile cells, each represented by a circle. The size of the circle gives the total population, and the color shows the percentage enslaved. This required modifying the usual county-level statistics with data from more than 150 cities and a handful of sub-county areas. With this new approach, cities now pop out from their rural surroundings, the much higher density of the North is clear, creating a time series for consistent comparisons is easy, and both relative and absolute population are visible at the same time.

One of my design dilemmas was finding a way to put the visual emphasis where it belonged—on the spread of slavery—while avoiding the usual tendency to associate population density with darker colors, which could carry a problematic association with race. A black background makes it possible to use bright, vivid colors for the most important areas; it also makes the overall mood of the map more somber, even sinister. Visually, slavery becomes a wildfire burning through the South.

This map prioritizes people over political boundaries. In this sense, the bubble grid is a kind of undistorted cartogram, sacrificing coastlines and county shapes to make room for a cleaner, more incisive treatment of population.

We stand atop the mountains and feel the pull of the landscape.

North Cascades Collaboration »

Jason Curtis and Nikki Frumkin

As our story begins, a watercolorist and a carto-photographer, both from the Pacific Northwest, meet on a rugged mountain peak—metaphorically. In this collaborative work, we interpreted the vista via our creative specialties, bringing two points of view and years of experience to the task.

Watercolorist Nikki Frumkin is often found in the wild with her sketchbook, painting summits, ridges, and alpine lakes. Her works combine pattern, vibrant color, and energetic lines to capture and share the feeling of being in the mountains. When sketching for this piece in situ, she often had to make room for curious mountain goats to wander through her workspace.

Carto-photographer Jason Curtis brought his own version of the "name-that-peak" game to his medium. His creations combine spectacular views with GIS-enabled, superimposed labels, giving each peak its rightful name and bringing the viewer more in touch with the mountains. This approach meets the audience at eye level, our natural perspective. For some viewers, the pieces are a framing of fond memories; for others, they are a to-do list of future adventures; and for still others, they are simply organized beauty.

To retain legibility when placing words on watercolor, Curtis varied the opacity along each leader line to allow the watercolor to come through as much as possible. The foreground labels posed an extra challenge because of the darker, textured glacier and rock behind them. Rather than applying a solid color overlay or blurring the background, Curtis placed the labels on cards pulled from the sky, allowing the two foreground labels to reside "in the sky" with the other peak labels, and adding to the sense of depth.

We both approach cartography from what is now an unusual point of view, but one that was common historically: we are mountaineers as well as artists. We stand atop the mountains and feel the pull of the landscape. To any mountain climber who has achieved a summit, the vista isn't just a prize. The view is a panorama of invitation, a checklist of future objectives, and an unlimited source of wonder and adventure. Our artwork draws viewers into an alien world of mountaintops and thin air, as though we are all journeying toward the next peak together.

Learn more at jasoncurtisphotography.com and drawntohighplaces.com

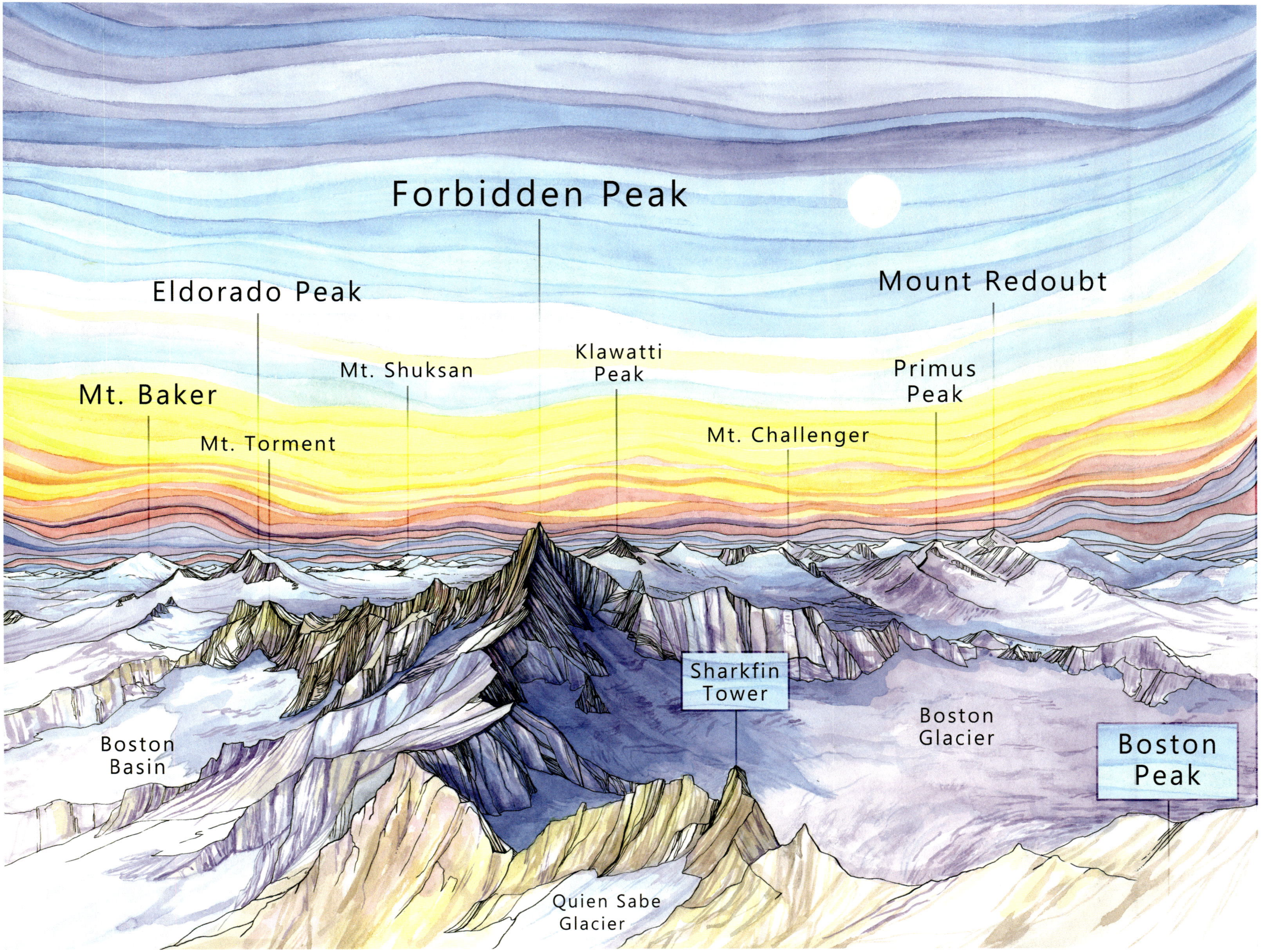
Forbidden Peak
Eldorado Peak
Mount Redoubt
Klawatti Peak
Mt. Shuksan
Primus Peak
Mt. Baker
Mt. Torment
Mt. Challenger
Sharkfin Tower
Boston Glacier
Boston Basin
Boston Peak
Quien Sabe Glacier

Famine in Yemen »

Laris Karklis — The Washington Post

Over the course of one month, I worked with colleagues in the Foreign Department at *The Washington Post* on a series of maps explaining how famine conditions in Yemen are being driven by human conflict rather than by natural causes such as drought. We agreed that a map-driven page was the most effective way to communicate this story.

I wanted readers to consider three critical questions:

- Why is 60 percent of the Yemeni population in urgent need of humanitarian assistance?
- How is food insecurity connected to the ability to deliver aid?
- Which forces in the civil war are creating this massive internal displacement?

At the *Post*, we present our work to the team in three stages before publication: an initial pitch with broad outlines of the concept, a review of the development of the specifics of maps and graphics, and a final look at the flow and interplay of the maps and the narrative.

The Yemen map includes five overlapping thematic datasets: conflict zones, food insecurity status, road conditions, forced migrations, and territorial control. The most significant design decision I made was limiting my palette to tones of purple, red, and orange, rather than choosing separate colors for each feature. Red and purple are not easy to combine because they are so closely related. Because the famine areas are present throughout the entire country, the red to orange blend used to depict conflict zones contrasts well enough to distinguish those areas from the overall purple hue. The vector linework continues the same interplay with roads classified in red and orange and the displacement sites shown in purple. Not wanting to add a fourth color, I chose gray dashes and a subtle hatch to represent territorial control.

I hope that this map helps to convey the tragic results of a civil war that has decimated Yemen's economy and brought joblessness, suffering, and starvation to millions.

See full Washington Post article at wapo.st/2ok8W6B

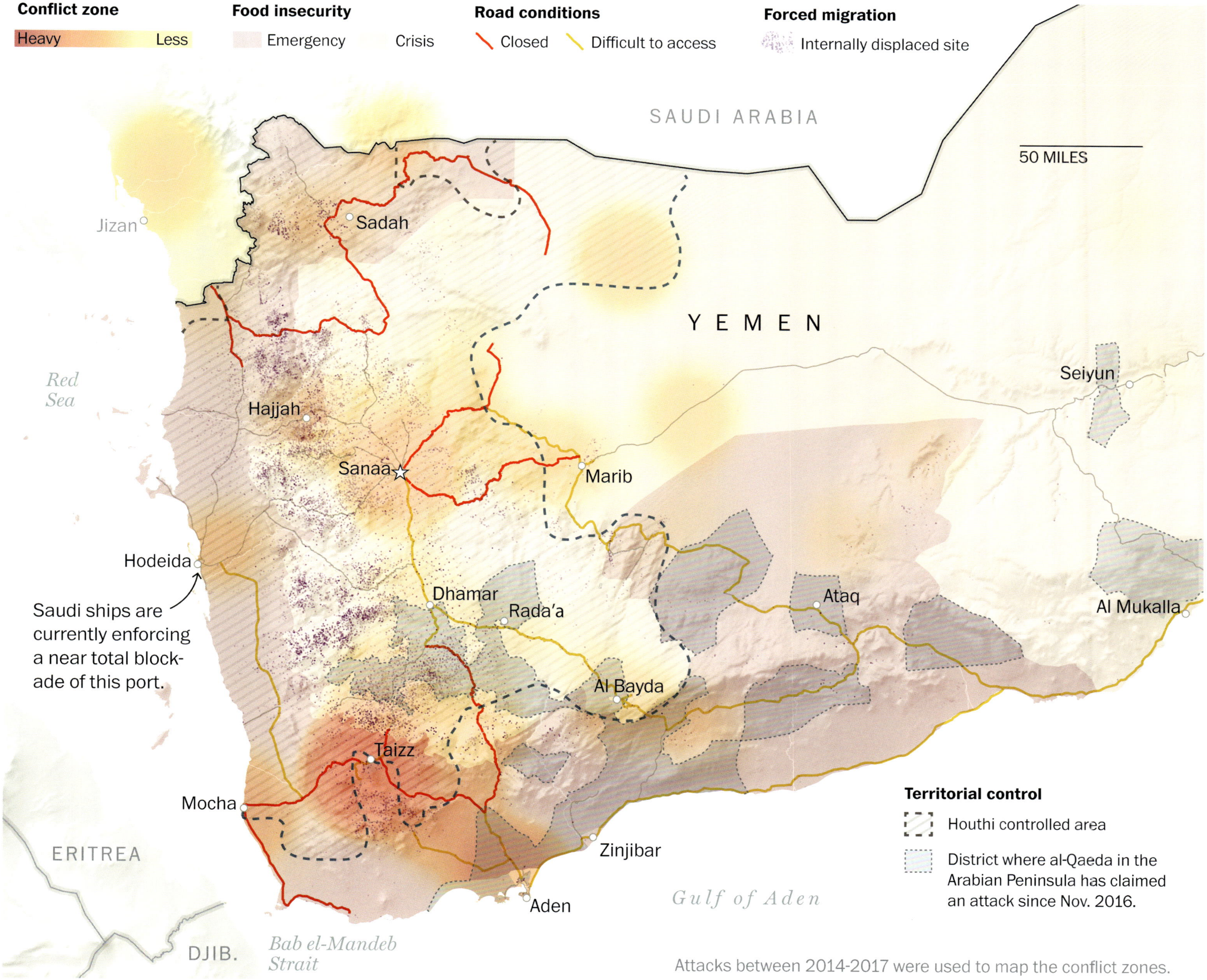
Conflict zone
Heavy
Less
Food insecurity
Emergency
Crisis
Road conditions
Closed
Difficult to access
Forced migration
Internally displaced site
SAUDI ARABIA
50 MILES
Jizan
Sadah
YEMEN
Red Sea
Hajjah
Sanaa
Seiyun
Marib
Hodeida
Saudi ships are currently enforcing a near total block-ade of this port.
Dhamar
Rada'a
Ataq
Al Mukalla
Al Bayda
Taizz
Mocha
ERITREA
Zinjibar
Aden
Gulf of Aden
DJIB.
Bab el-Mandeb Strait
Territorial control
Houthi controlled area
District where al-Qaeda in the Arabian Peninsula has claimed an attack since Nov. 2016.
Attacks between 2014-2017 were used to map the conflict zones.

The stars were right in front of our eyes the whole time, but were invisible, just as they are in the actual sky.

(In)Visible Night Sky »

Martin Pulicar

Billions of stars light up the sky every night. Many of them shine with much greater intensity than our sun. Nearly all these stars are so far away that our eyes cannot see them, but each star is still there. How is it that we stare directly at them, yet they are invisible?

This map shows some 5,000 stars observable with the naked eye. Nothing new, you say? Exactly right. Most night sky maps show us what we already know, what we already can see. But illuminate this map with ultraviolet (UV) light, and it reveals a sea of tiny dots—the exact locations of another 250,000 fainter stars in our cosmic vicinity. The stars were right in front of our eyes the whole time, but were invisible, just as they are in the actual sky.

My offset and UV-sensitive silk screen combination map has the bold goal of expanding each viewer's perspective. Printed maps are rarely interactive, so breaking this rule with UV light also broadens the horizons of cartography a bit.

But the primary focus of my map is somewhere else.

Even though the map looks extremely crowded under the UV light, it still reveals only a *tiny fraction* of what is out there in our galaxy. In fact, there are still roughly *400,000 times more stars* in the Milky Way than are shown on this map. And our galaxy is only one of hundreds of billions of galaxies. Everyone we have ever known; everything humans have ever created, touched, or destroyed; and all the worlds we have ever visited and mapped are confined to a few grains of sand orbiting an average star somewhere in the suburbs of a mediocre galaxy. We should not forget these humbling facts.

I hope this map will encourage viewers to preserve our fragile little blue spaceship and embark on journeys to explore as many new horizons as possible. So much is out there, still invisible.

Learn more at pulicar.xyz

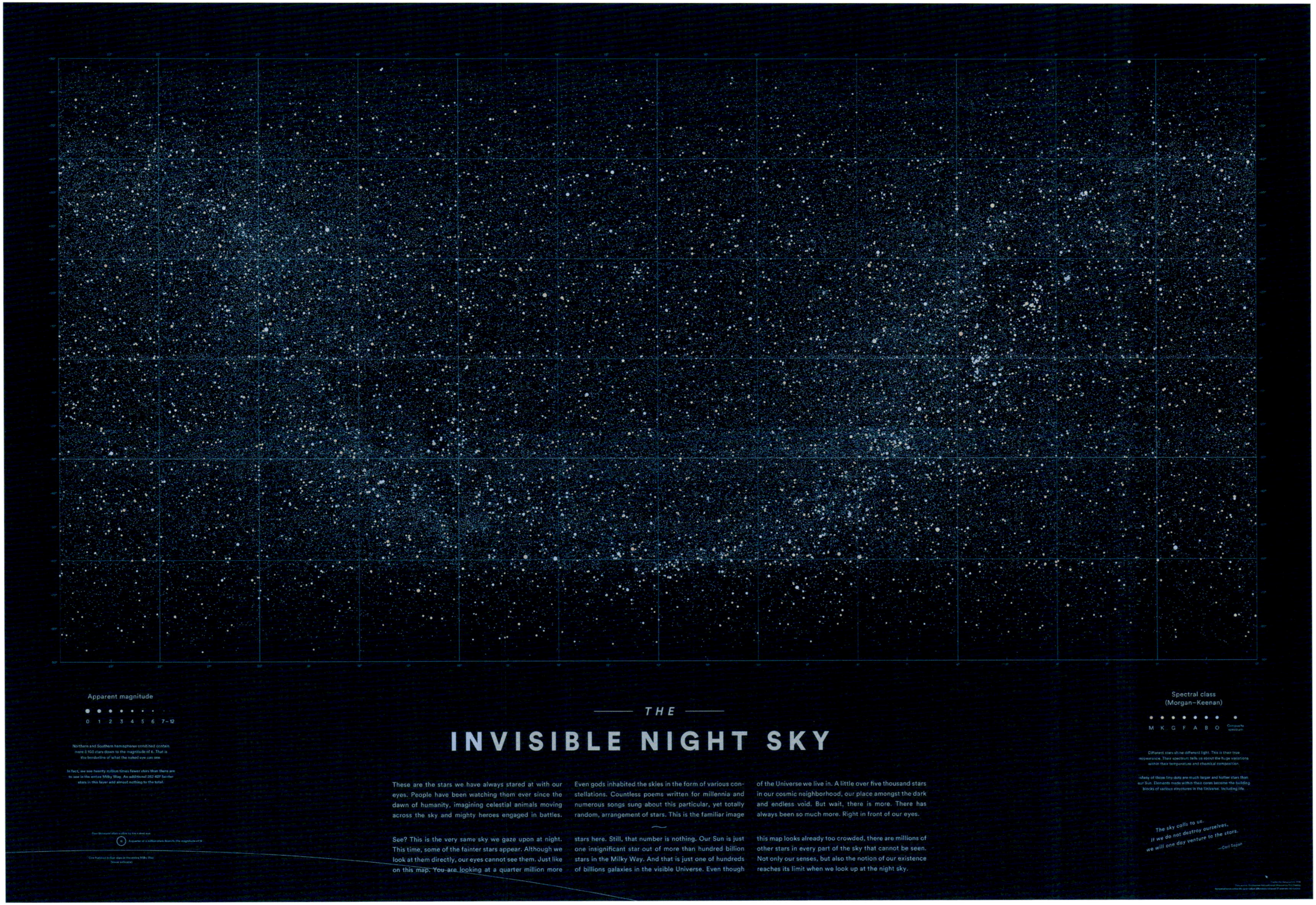
Apparent magnitude
0 1 2 3 4 5 6 7–12
Northern and Southern hemispheres combined contain mere 5 105 stars down to the magnitude of 6. That is the borderline of what the naked eye can see.
In fact, we see twenty million times fewer stars than there are to see in the entire Milky Way. An additional 252 427 fainter stars in this layer add almost nothing to the total.
THE
INVISIBLE NIGHT SKY
These are the stars we have always stared at with our eyes. People have been watching them ever since the dawn of humanity, imagining celestial animals moving across the sky and mighty heroes engaged in battles. Even gods inhabited the skies in the form of various constellations. Countless poems written for millennia and numerous songs sung about this particular, yet totally random, arrangement of stars. This is the familiar image of the Universe we live in. A little over five thousand stars in our cosmic neighborhood, our place amongst the dark and endless void. But wait, there is more. There has always been so much more. Right in front of our eyes.
See? This is the very same sky we gaze upon at night. This time, some of the fainter stars appear. Although we look at them directly, our eyes cannot see them. Just like on this map. You are looking at a quarter million more stars here. Still, that number is nothing. Our Sun is just one insignificant star out of more than hundred billion stars in the Milky Way. And that is just one of hundreds of billions galaxies in the visible Universe. Even though this map looks already too crowded, there are millions of other stars in every part of the sky that cannot be seen. Not only our senses, but also the notion of our existence reaches its limit when we look up at the night sky.
Spectral class
(Morgan–Keenan)
M K G F A B O
Different stars shine different light. This is their true appearance. Their spectrum tells us about the huge variations within their temperature and chemical composition.
Many of those tiny dots are much larger and hotter stars than our Sun. Elements made within their cores become the building blocks of various structures in the Universe. Including life.
The sky calls to us.
If we do not destroy ourselves,
we will one day venture to the stars.
—Carl Sagan

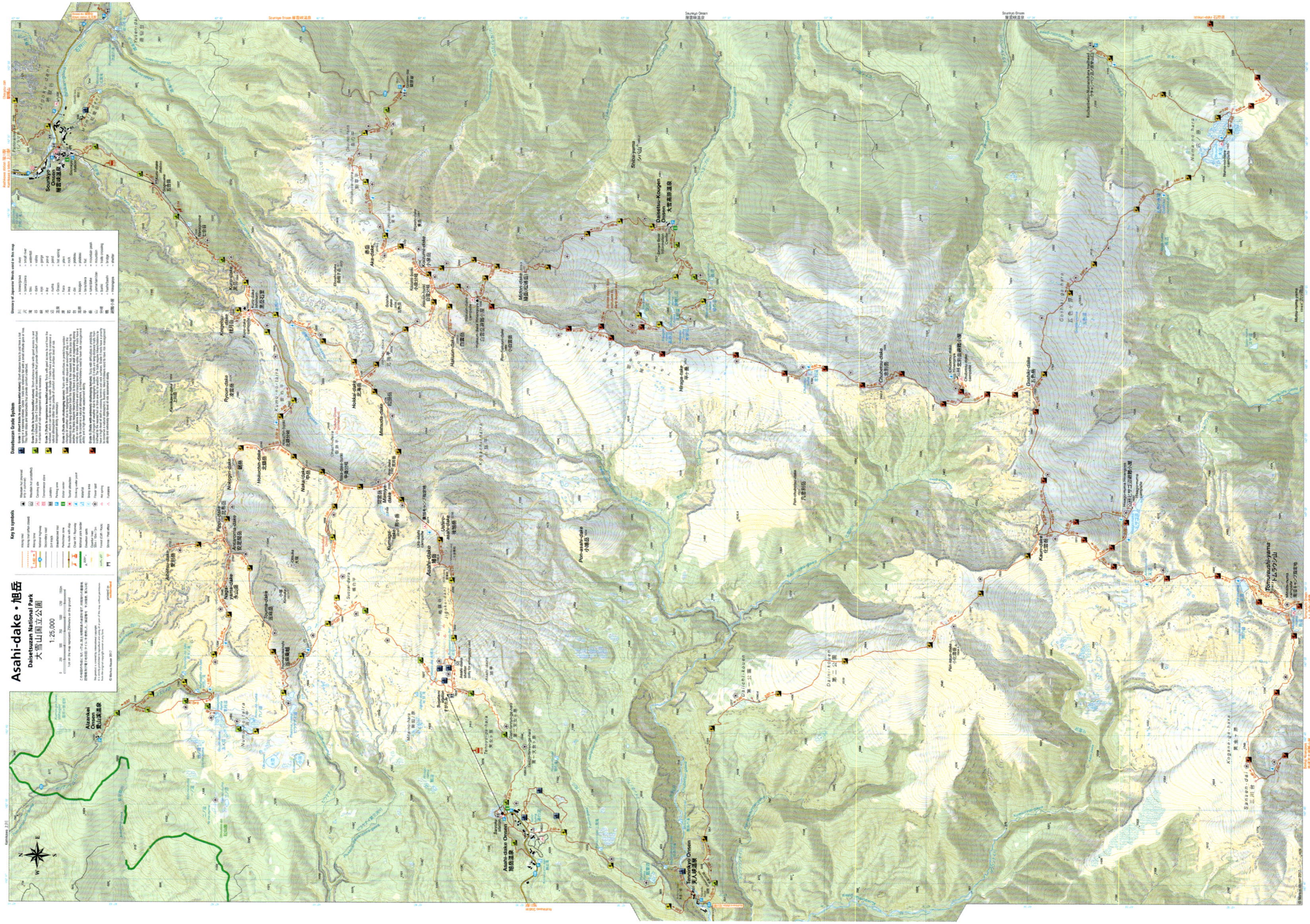
Asahi-dake • 旭岳
Daisetsuzan National Park
大雪山国立公園
1:25,000

« Asahi-dake, The Heart of the Daisetsuzan National Park

Markus Hauser—Orell Füssli Kartographie AG

My life has been punctuated by a fascination with the art of master cartographer Eduard Imhof. As a schoolboy, I copied his colorful maps, and I began my training as a cartographer with Orell Füssli in 1986, within a month of Imhof's death. I was fortunate to learn without the aid of computers, implementing Imhof's approach to the harmonious relationship of lines, patterns, and colors. Now I work for the company with which Imhof had a strong working relationship from his early years at the ETH Zurich.

Today, when small mobile phone screens and free maps dominate our competitive marketplace, applying my specific knowledge and experience isn't always easy. Fortunately, there are still places where tradition plays an important role in mapmaking and people attach a great deal of importance to craftsmanship.

In Japan, my Swiss training and skills in creating topographic maps find fertile ground as increasing numbers of foreign tourists need clear, understandable maps for hiking. Through the vision and with the help of Teiji Watanabe from the University of Hokkaido, I created a bilingual map of Mount Asahi-dake and its surroundings in the heart of the Daisetsuzan National Park on Hokkaido. I incorporated elements of Japanese cartographic tradition, including special map symbols such as the one for *onsen*, a hot spring. Our map is also the first to present the new classification system for hiking trails developed by the Japanese Ministry of the Environment, known as "Daisetsuzan Grade."

In this map, I achieved a synthesis of modern technology and European and Japanese cartographic traditions using Japanese base data, hand-drawn digital shaded relief by German artist Gunter Kaiser, and the Swiss-style map design I have spent a lifetime learning. Teiji Watanabe and his team provided expert contributions and editing.

The extensive representation of vegetation on hiking maps is unusual in Japan. Against this background, the representation of the forest on the map is an important innovation, which makes it much more user-friendly for hikers. This, however, did not allow me to use Imhof's preferred hypsometric color schema. But as the crowning glory and appreciation of Imhof, I added a pink shimmer on the highest peaks in the land of the rising sun.

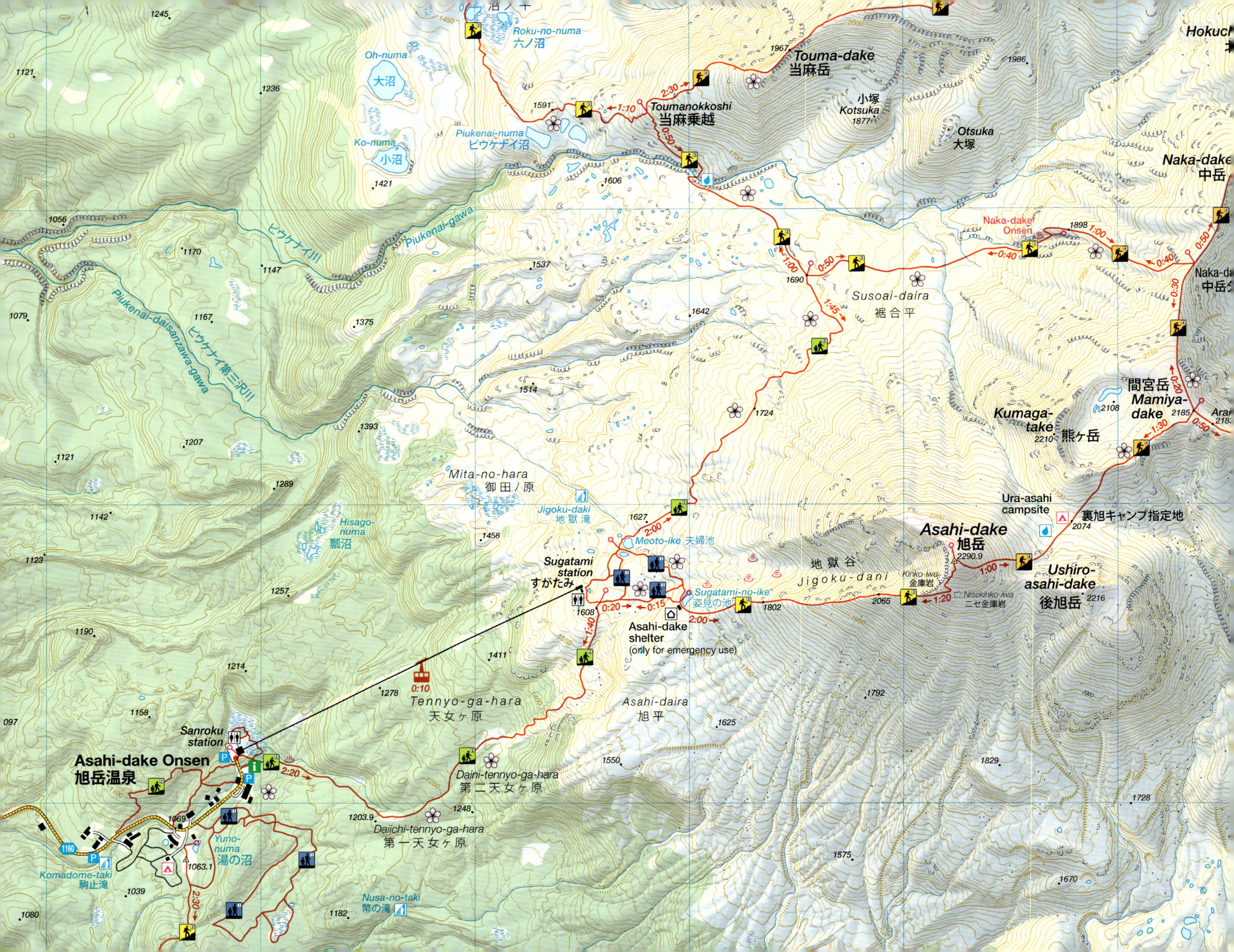

Roku-no-numa
六ノ沼
Oh-numa
大沼
Ko-numa
小沼
Piukenai-numa
ピウケナイ沼
Toumanokkoshi
当麻乗越
Touma-dake
当麻岳
小塚
Kotsuka
1877
Otsuka
大塚
Hokuch
Naka-dake
中岳
Naka-dake Onsen
Naka-
中岳
Susoai-daira
裾合平
ピウケナイ川
Piukenai-gawa
Piukenai-daisanzawa-gawa
ピウケナイ第三沢川
Mita-no-hara
御田ノ原
Jigoku-daki
地獄滝
Hisago-numa
瓢沼
Meoto-ike
夫婦池
Sugatami station
すがたみ
Sugatami-no-ike
姿見の池
Asahi-dake shelter
(only for emergency use)
地獄谷
Jigoku-dani
Kinko-iwa
金庫岩
Nisekinko-iwa
ニセ金庫岩
Asahi-dake
旭岳
2290.9
Ura-asahi campsite
裏旭キャンプ指定地
2074
Ushiro-asahi-dake
後旭岳
2216
Kumaga-take
熊ヶ岳
2210
間宮岳
Mamiya-dake
2185
2108
Ara
Tennyo-ga-hara
天女ヶ原
Asahi-daira
旭平
Sanroku station
Asahi-dake Onsen
旭岳温泉
Daini-tennyo-ga-hara
第二天女ヶ原
Daiichi-tennyo-ga-hara
第一天女ヶ原
Yuno-numa
湯の沼
Komadome-taki
駒止滝
Nusa-no-taki
幣の滝
1160
1063.1
1069
1203.9
1248
1245
1121
1236
1967
1986
1591
1606
1421
1056
1170
1147
1537
1690
1898
1079
1167
1375
1642
1724
1514
1393
1207
1121
1289
1142
1627
1458
1123
1257
1608
1802
2065
1190
1214
1411
1278
1158
1792
1625
1550
1829
1728
1575
1670
1039
1080
1182
097
0:10
2:30
1:10
0:50
1:00
0:50
1:45
0:40
1:00
0:40
0:50
0:30
0:20
1:30
0:50
2:00
0:20
0:15
2:00
1:40
1:00
1:20
2:20
2:30

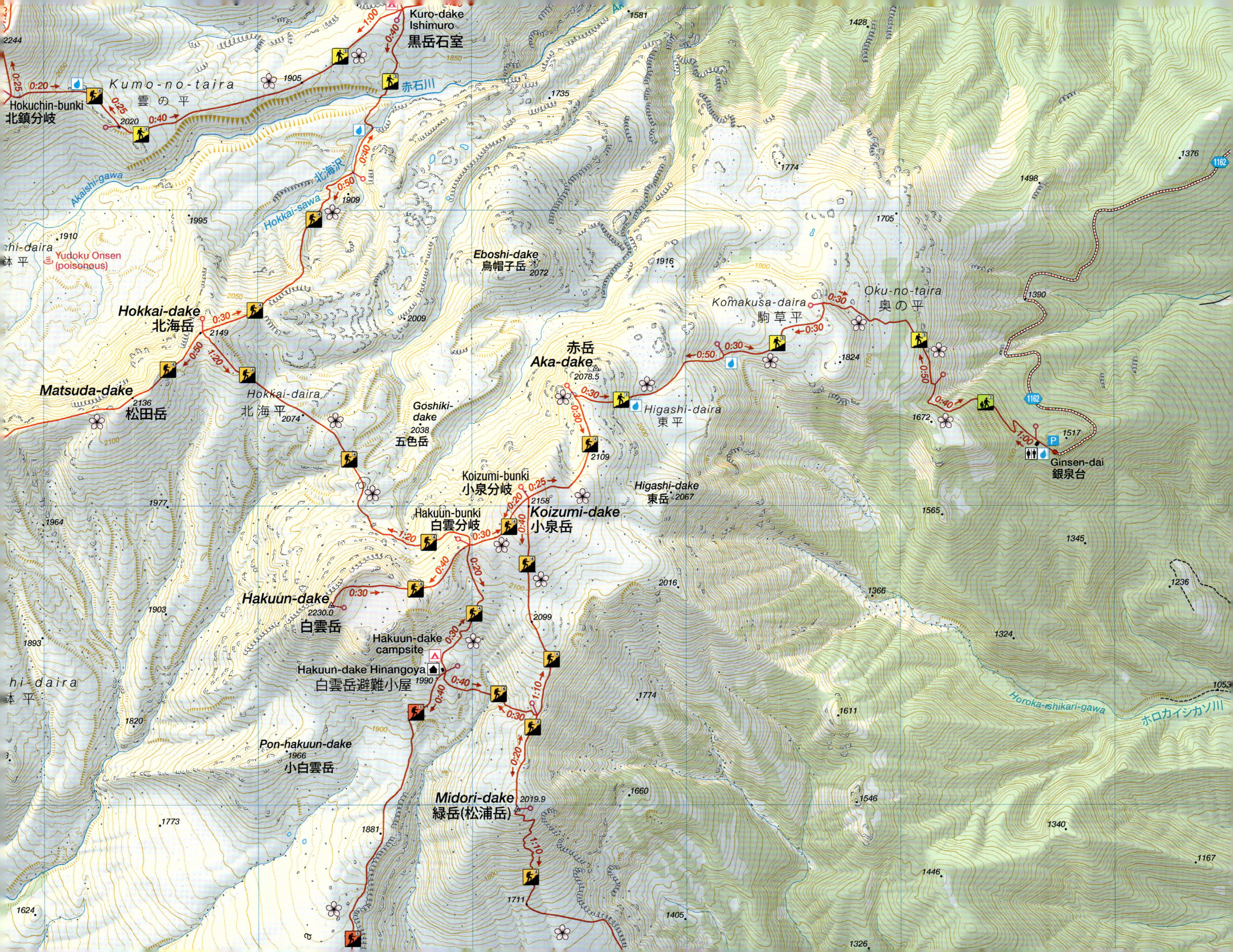

Kuro-dake Ishimuro
黒岳石室
Kumo-no-taira
雲の平
Hokuchin-bunki
北鎮分岐
赤石川
Akaishi-gawa
北海沢
Hokkai-sawa
Yudoku Onsen (poisonous)
Eboshi-dake
烏帽子岳
Hokkai-dake
北海岳
Matsuda-dake
松田岳
Hokkai-daira
北海平
Goshiki-dake
五色岳
赤岳
Aka-dake
Komakusa-daira
駒草平
Oku-no-taira
奥の平
Higashi-daira
東平
Ginsen-dai
銀泉台
Koizumi-bunki
小泉分岐
Koizumi-dake
小泉岳
Higashi-dake
東岳
Hakuun-bunki
白雲分岐
Hakuun-dake
白雲岳
Hakuun-dake campsite
Hakuun-dake Hinangoya
白雲岳避難小屋
Pon-hakuun-dake
小白雲岳
Midori-dake
緑岳(松浦岳)
Horoka-ishikari-gawa
ホロカイシカリ川

How Deep is Dark Star? »

Charles Preppernau, Martin Gamache, Scott Zillmer, and Michael Fry — National Geographic Magazine

"How Deep Is Dark Star" was created for a *National Geographic* magazine story about a remote mountain cave system in Uzbekistan with entrances high in the cliffs following the tilt of rock layers. The map had to show the cave structure clearly while also showing its relationship to the surrounding terrain and geology. We wanted to emphasize the difficulty of reaching the entrances and the caves' overall isolation. Unfortunately, the caves, cliffs, and mountains are arranged so that no single view can capture them in a way that clearly shows their relationship to each other.

It's often the case that visualizing a subject in a realistic way doesn't quite capture it. The solution to this problem is to find an acceptable break with reality that works. We decided to cut away the intervening rock and draw the cliff walls as if looking at them from inside the mountain to show where the caves are in relation to both the geology and the cliffs. The parts of the cliff walls visible from the outside have a distinct texture, and they serve as a natural legend that the viewer can link to the cutaway geology.

Initially, we expected the cavern survey data to be the hardest part to include in this 3D scene, but in fact, the terrain required the most work. Much of the area is in a cluster of bad elevation data and had to be reconstructed from Soviet-era topographic maps, then sculpted in a 3D application to match the profile and sharpness of the cliffs seen in photos from an expedition team. The camera view that worked for the cave and cliffs was from a high angle, making the surrounding hills appear taller in the background. We again broke with reality by bending the terrain in the background downward.

By making several carefully chosen breaks with reality, we brought the Boysuntov Range into clearer focus. We were able to give our audience a true sense of how high and remote these mountain caves really are.

How Deep Is Dark Star?

Uzbekistan's remote Boysuntov Range is hard to access, and the region is politically unstable. Still, the mysteries of Dark Star and nearby Festivalnaya—two of the world's deepest high-altitude cave systems—are an enduring lure for explorers.

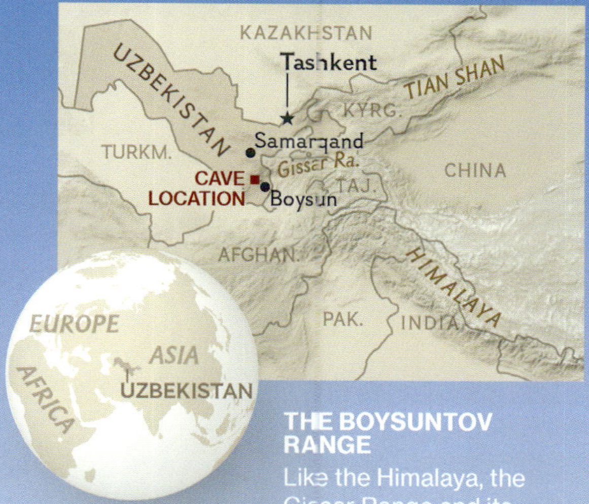

THE BOYSUNTOV RANGE

Like the Himalaya, the Gissar Range and its cave-rich spur, the Boysuntov (illustrated below), were forced upward by the collision of Indian and Eurasian plates.

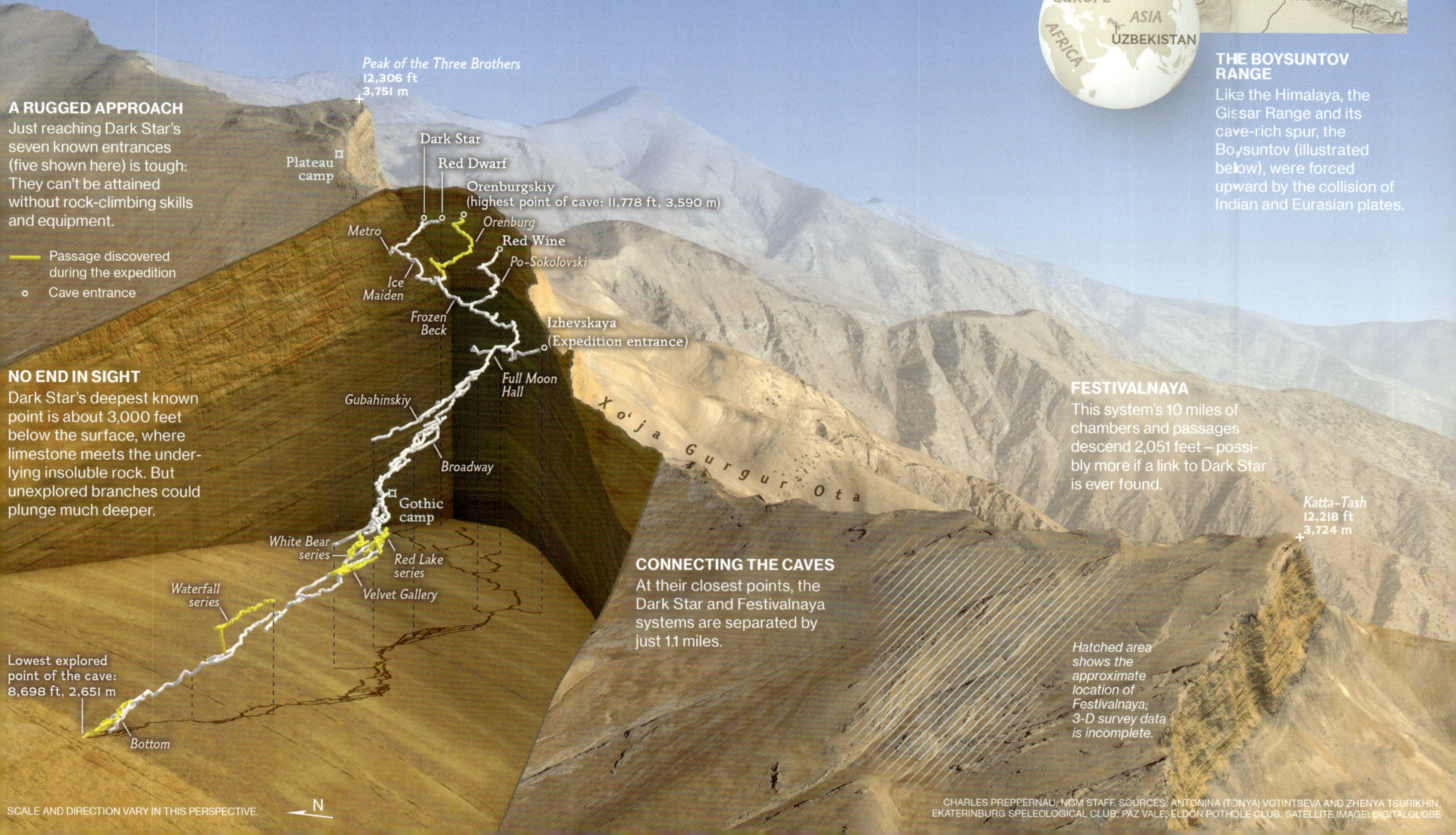

A RUGGED APPROACH

Just reaching Dark Star's seven known entrances (five shown here) is tough: They can't be attained without rock-climbing skills and equipment.

NO END IN SIGHT

Dark Star's deepest known point is about 3,000 feet below the surface, where limestone meets the underlying insoluble rock. But unexplored branches could plunge much deeper.

CONNECTING THE CAVES

At their closest points, the Dark Star and Festivalnaya systems are separated by just 1.1 miles.

FESTIVALNAYA

This system's 10 miles of chambers and passages descend 2,051 feet – possibly more if a link to Dark Star is ever found.

CHARLES PREPPERNAU, NGM STAFF. SOURCES: ANTONINA (TONYA) VOTINTSEVA AND ZHENYA TSURIKHIN, EKATERINBURG SPELEOLOGICAL CLUB; PAZ VALE, ELDON POTHOLE CLUB. SATELLITE IMAGE: DIGITALGLOBE

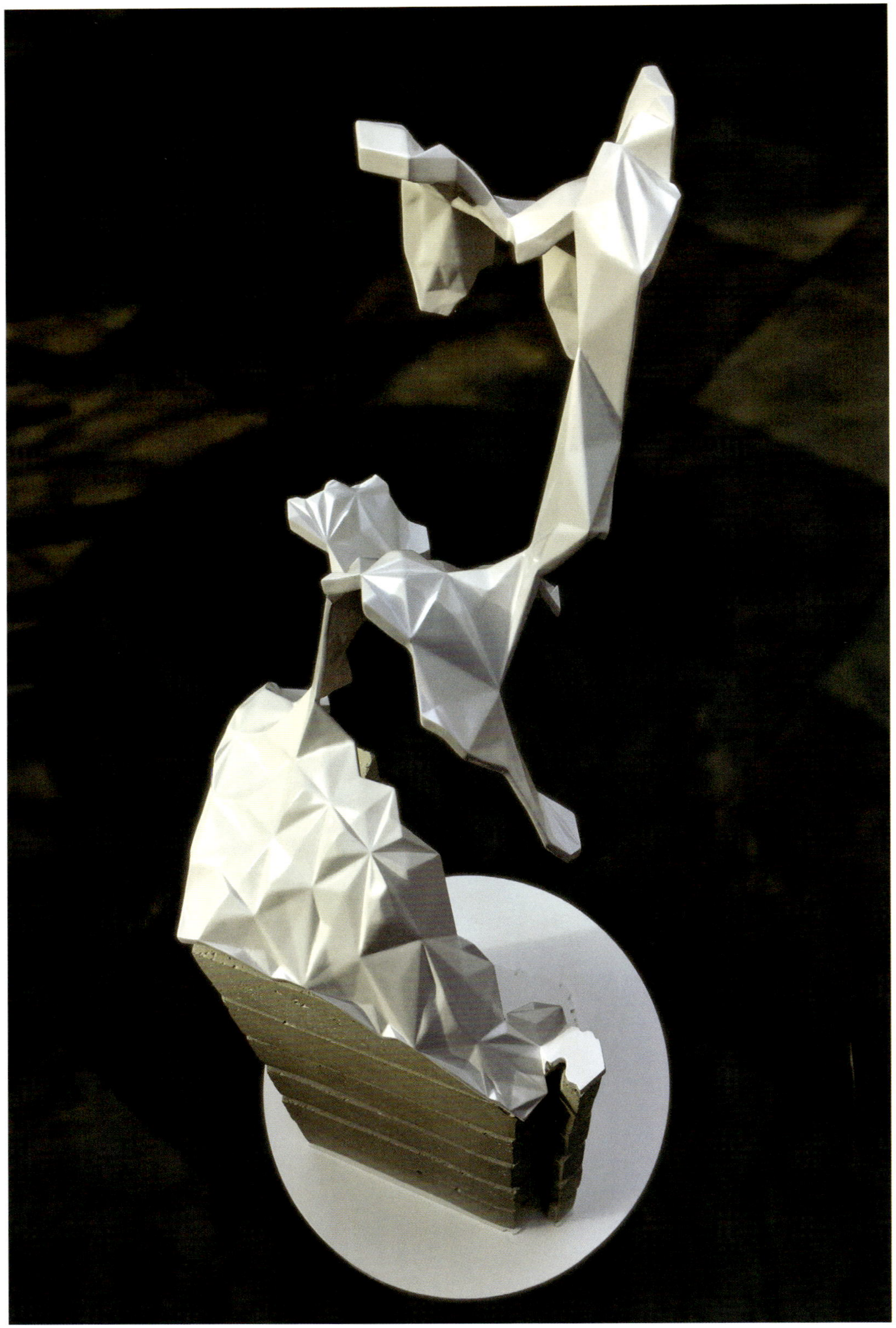

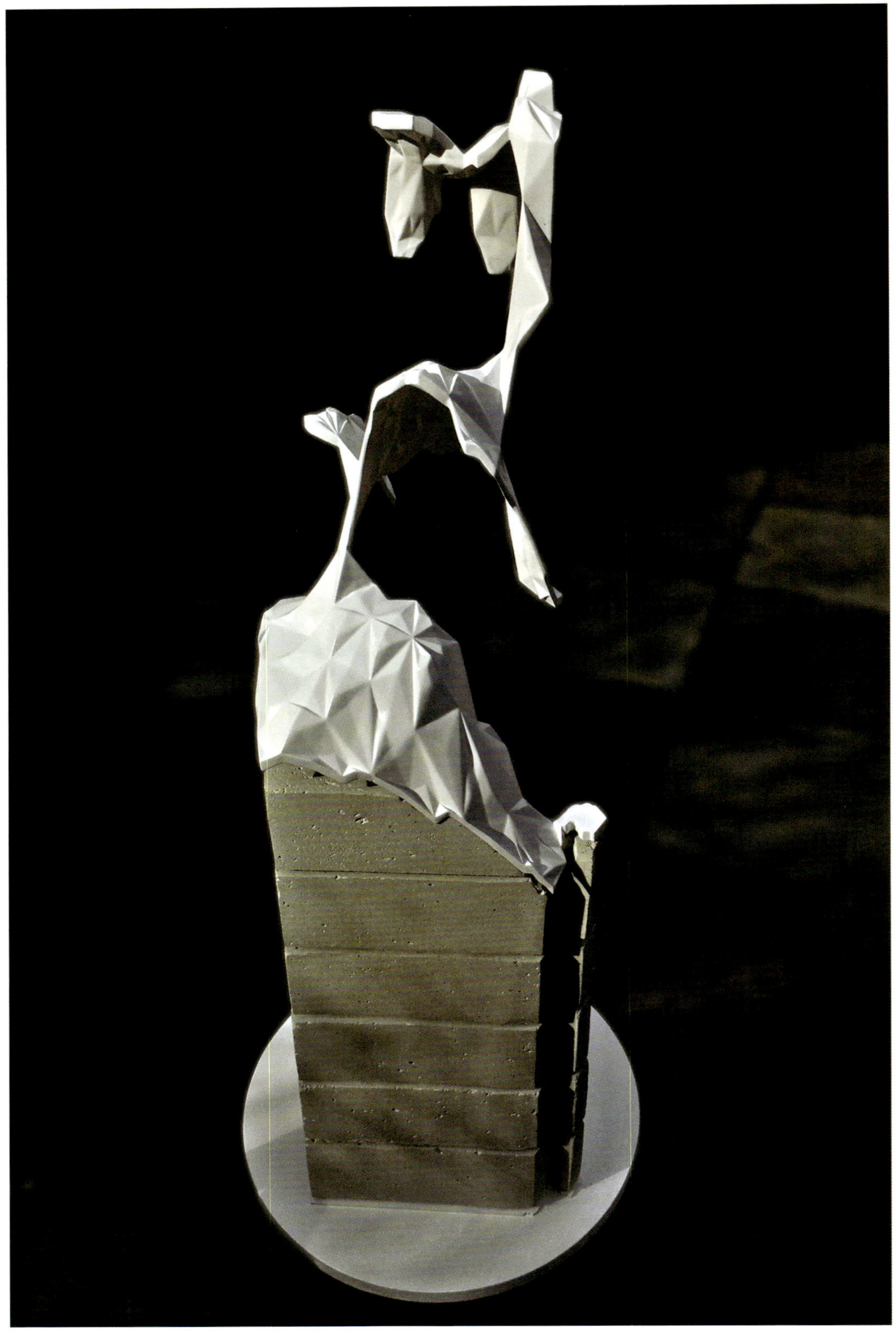

《 Housing Prices Ripping San Francisco Apart

Doug McCune

This data sculpture depicts housing prices in San Francisco, and it's the story of a city ripping apart at the seams. The height of the sculpture represents the average price per square foot of homes sold in each area. If neighboring areas are close in value, they are connected; if they are too far from each other in terms of home prices, they are split, tearing the city along its most extreme economic divides.

The data consists of transaction records of approximately 5,000 home sales. The transaction points are aggregated into hexagons to calculate the average price per square foot for each region, then turned into a 3D model by raising each region to a height proportional to its average home price. Neighboring areas with a large gap between their average prices are split apart, organically producing the spiral pattern. When looking at the sculpture from above, all its pieces line up to form a cohesive outline of San Francisco in a shape more identifiable as a map. But from the side, the city is ripped apart and stratified by class.

The differences among areas of the city are easy to see. The lowest per-square-foot home prices are in Bayview/Hunters Point in the southeast quadrant, and prices increase moving northwest. The general trend is that homes are more expensive the closer they are to the Marina, the Embarcadero, and the northeast edge of the city.

The concrete base of the sculpture serves as a map legend. Each vertical section of the base represents a price of $100 per square foot. From the bottom, the viewer can count up to reach a starting price of approximately $500-$600 per square foot, which is about the minimum cost of a home in the cheapest parts of San Francisco today.

This sculpture is my attempt, as a resident of the Bay Area, to come to grips with the insanity of the San Francisco housing market and the vast economic inequality in one of the most expensive places in the United States.

Learn more at dougmccune.com

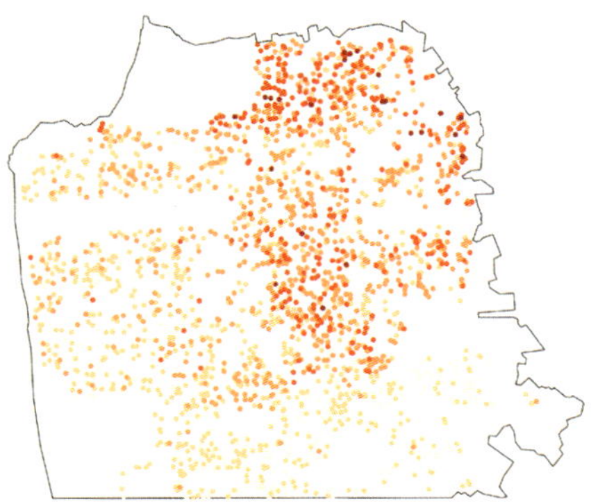

Point data from Redfin shows 5,000 recent home sales in San Francisco. Each transaction is color-coded by the price per square foot.

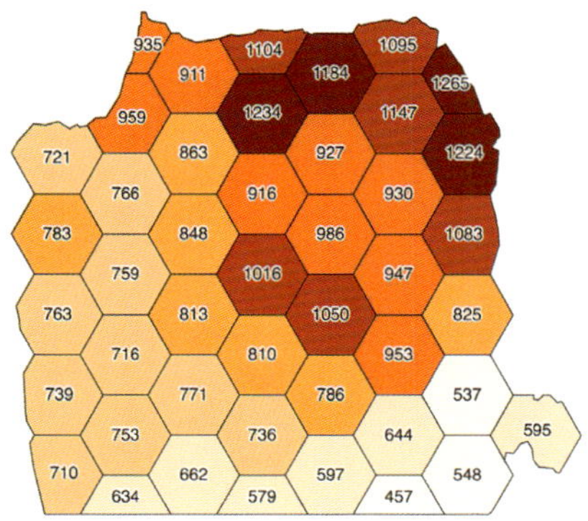

Data binned into hexagons shows average home prices per square foot.

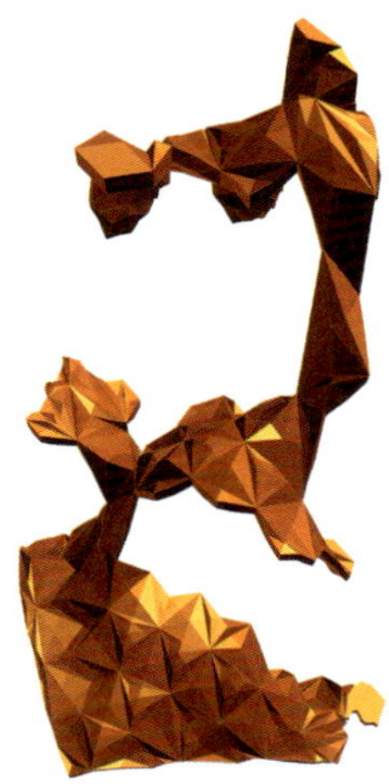

To create a 3D model, we determined a threshold for how close together prices in neighboring regions have to be to connect. Any neighboring regions that exceed that delta are split from each other, organically producing a spiral pattern through the city.

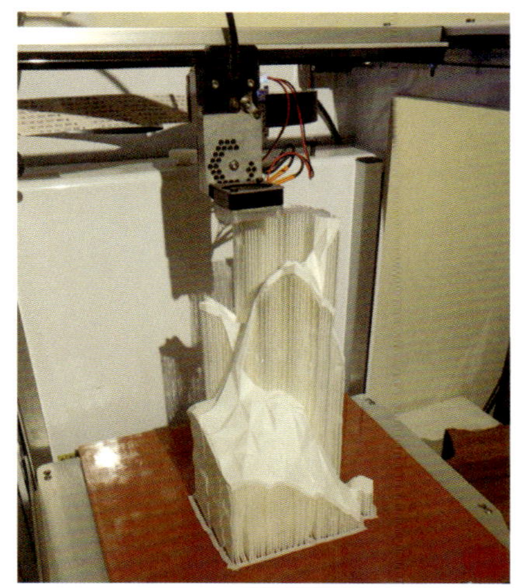

A small-scale prototype printed on a 3D printer preceded the final version. The final sculpture took over 90 hours to print. It was printed in seven large sections that were later pieced together.

El Reino de España, a Decorative Map of Spain »

Mike Hall

As a freelance illustrator, I specialize in map design and architectural illustration. I admire design that utilizes a restricted color palette and period-style elegant type and decorative embellishments. This map of Spain—my adopted home since 2014—is inspired by these qualities and combines my two specialties.

The style and composition is influenced by seventeenth-century decorative atlas maps, particularly those of the Dutch cartographer Joan Blaeu (1596–1673). Typical design elements of this period such as detailed city scenes on the left and right, shields (escudos) representing each autonomous community, vignette-style shading, an elaborate border, and a title cartouche are all featured here.

I created most of the design using Adobe Illustrator. The exceptions are the city scenes, which are pen-on-paper drawings scanned and colored in Photoshop, and the shaded relief of the mountain ranges, which is inspired by the subtle shading of nineteenth-century maps and was also painted in Photoshop. I developed a hierarchy of symbols and text label sizes to represent the populations of each settlement. The two typefaces used, Brandon Grotesque and Kursivschrift, are modern imitations of type commonly used in the late nineteenth and early twentieth centuries.

All titles and names on the map are written in Castilian Spanish, with alternative names in regional languages written below or alongside in parentheses. This approach maintains linguistic consistency across the design, although in certain regions, including Basque Country, Catalonia, and Galicia, the regional languages are officially used and widely spoken.

Many of my commissioned designs are modern in style, but this map and others are profoundly influenced by my love of antique cartography and prints from the early modern era through the early twentieth century.

Learn more at thisismikehall.com

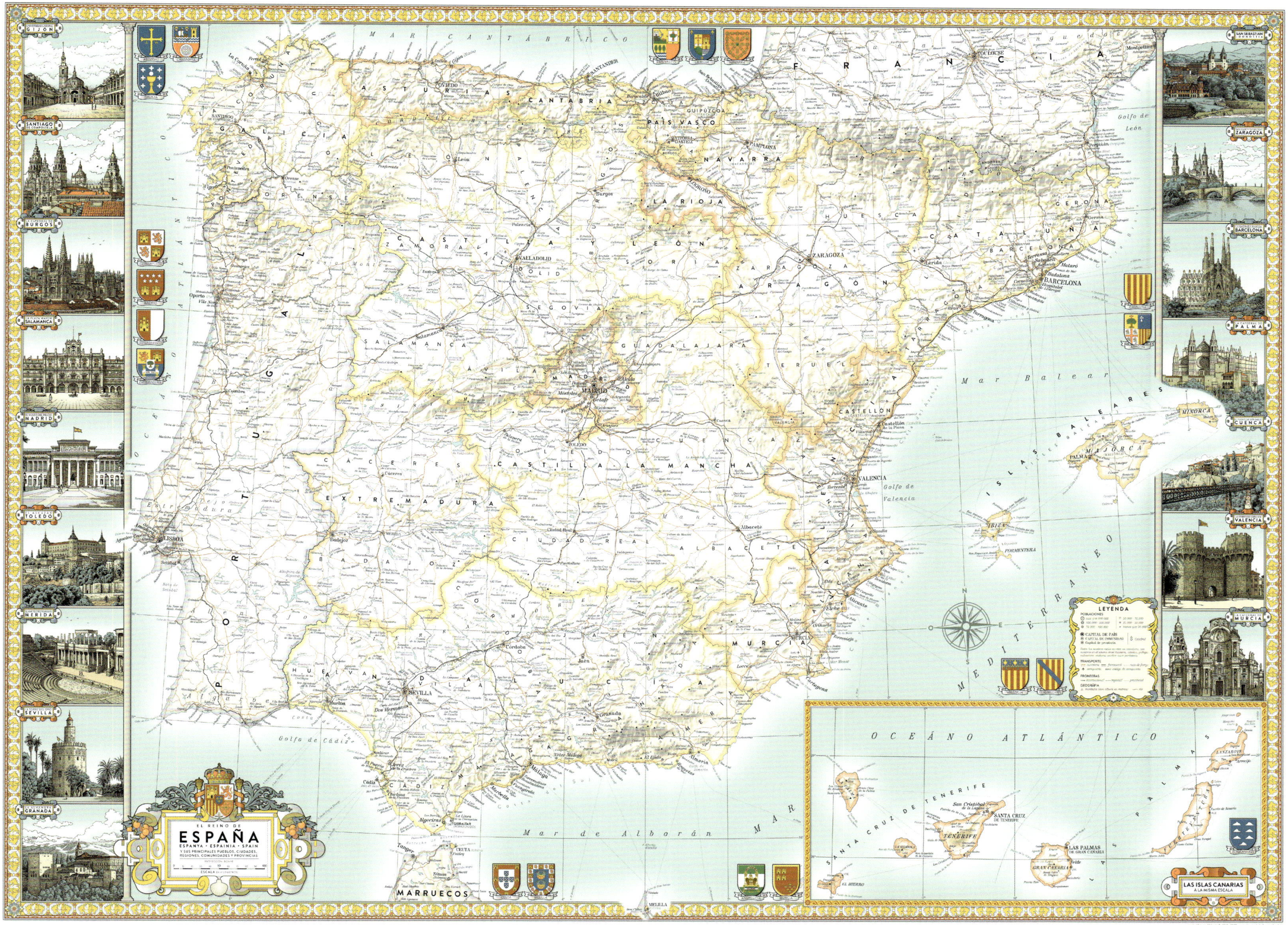
EL REINO DE
ESPAÑA
ESPANYA · ESPAINIA · SPAIN
Y SUS PRINCIPALES PUEBLOS, CIUDADES, REGIONES, COMUNIDADES Y PROVINCIAS
GIJÓN
SANTIAGO DE COMPOSTELA
BURGOS
SALAMANCA
MADRID
TOLEDO
MÉRIDA
SEVILLA
GRANADA
SAN SEBASTIÁN
ZARAGOZA
BARCELONA
PALMA
CUENCA
VALENCIA
MURCIA
MAR CANTÁBRICO
FRANCIA
ASTURIAS
CANTABRIA
PAÍS VASCO
NAVARRA
LA RIOJA
GALICIA
CASTILLA Y LEÓN
ARAGÓN
CATALUÑA
CASTILLA-LA MANCHA
EXTREMADURA
ANDALUCÍA
MURCIA
PORTUGAL
OCÉANO ATLÁNTICO
Golfo de León
Mar Balear
MEDITERRÁNEO
Golfo de Valencia
Golfo de Cádiz
Mar de Alborán
MARRUECOS
ISLAS BALEARES
MALLORCA
MENORCA
IBIZA
FORMENTERA
MADRID
BARCELONA
VALENCIA
ZARAGOZA
SEVILLA
LISBOA
LEYENDA
OCÉANO ATLÁNTICO
SANTA CRUZ DE TENERIFE
LAS PALMAS
TENERIFE
GRAN CANARIA
LANZAROTE
FUERTEVENTURA
EL HIERRO
LAS ISLAS CANARIAS
A LA MISMA ESCALA

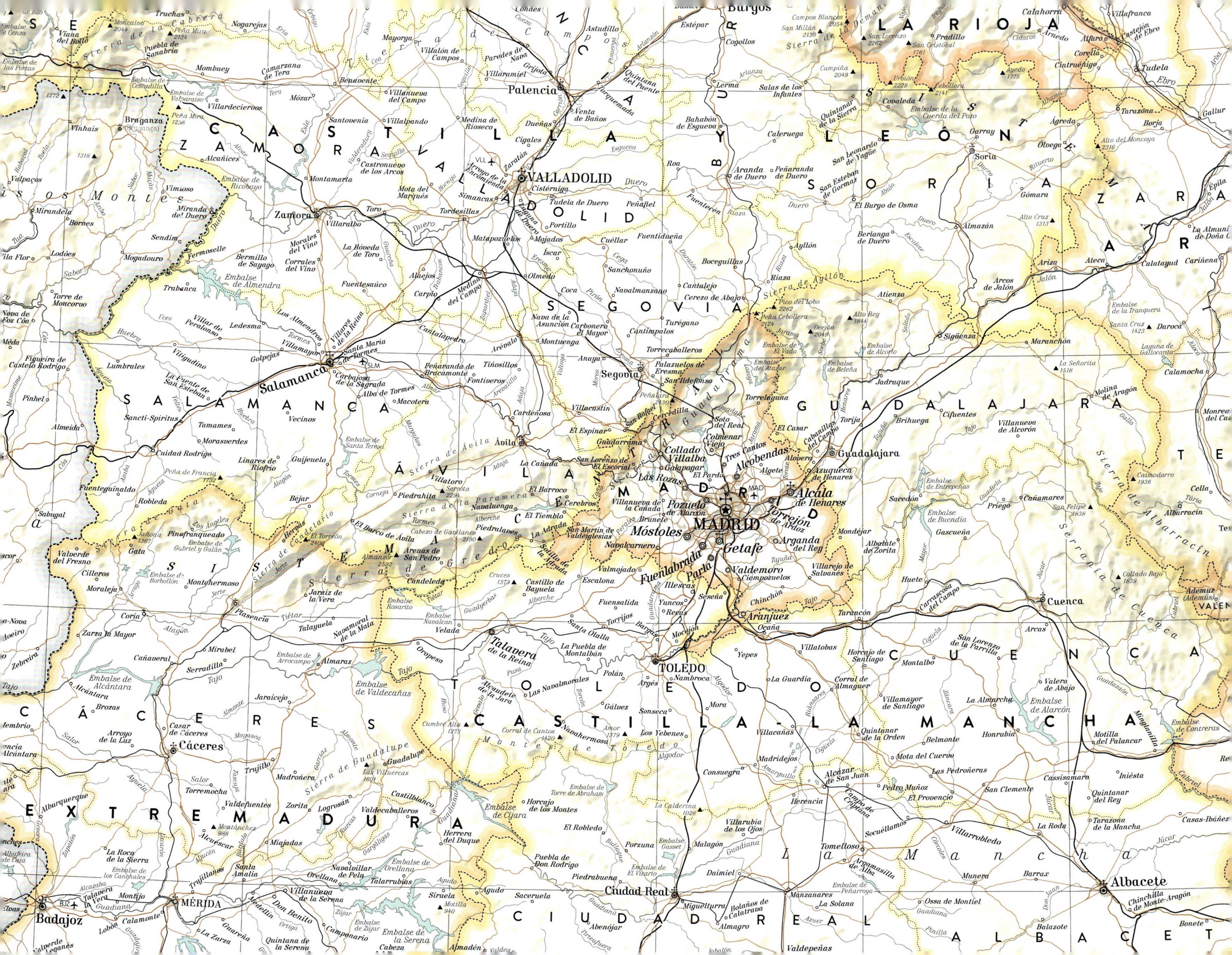

CASTILLA Y LEÓN
LA RIOJA
ZAMORA
VALLADOLID
SEGOVIA
SORIA
SALAMANCA
ÁVILA
MADRID
GUADALAJARA
ZARAGOZA
TERUEL
CUENCA
TOLEDO
CASTILLA-LA MANCHA
CÁCERES
EXTREMADURA
CIUDAD REAL
ALBACETE
BURGOS
Burgos
Palencia
Valladolid
Zamora
Salamanca
Segovia
Ávila
Soria
Guadalajara
Madrid
Toledo
Cuenca
Cáceres
Mérida
Badajoz
Ciudad Real
Albacete
Plasencia
Talavera de la Reina
Aranjuez
Alcalá de Henares
Móstoles
Getafe
Fuenlabrada
Parla
Alcobendas
Torrejón de Ardoz
Pozuelo de Alarcón
Collado Villalba
Las Rozas
Valdemoro
Tres Cantos
Medina del Campo
Tordesillas
Benavente
Braganza
Tudela
Calatayud
Sigüenza
Aranda de Duero
El Burgo de Osma
Almazán
Béjar
Coria
Trujillo
Guadalupe
Don Benito
Villanueva de la Serena
Puertollano
Manzanares
Tomelloso
Alcázar de San Juan
Villarrobledo
La Roda
Chinchilla de Monte-Aragón
Sierra de Guadarrama
Sierra de Gredos
Sierra de Gata
Sierra de Ayllón
Sierra de la Demanda
Serranía de Cuenca
Sierra de Albarracín
Montes de Toledo
Sierra de Guadalupe
La Mancha
Duero
Tajo
Ebro
Guadiana
Jarama
Tormes
Alberche
Embalse de Alcántara
Embalse de Valdecañas
Embalse de Entrepeñas
Embalse de Buendía
Embalse de Alarcón
Embalse de Cíjara
Embalse de Almendra
Embalse de Ricobayo

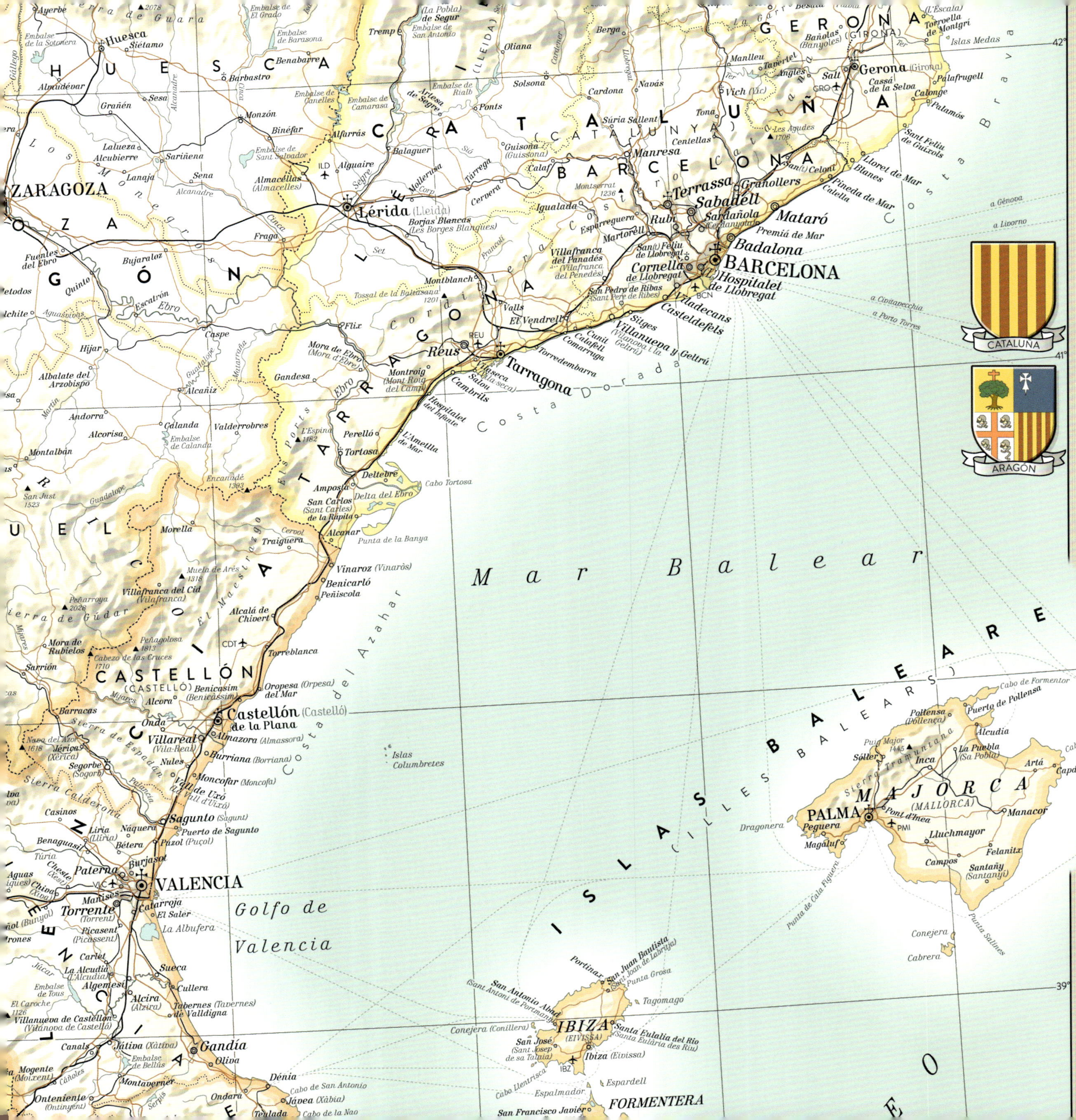

GERONA
Gerona (Girona)
HUESCA
Huesca
CATALUÑA (CATALUNYA)
BARCELONA
BARCELONA
Terrassa
Sabadell
Mataró
Badalona
Hospitalet de Llobregat
Cornellà de Llobregat
Granollers
Manresa
Lérida (Lleida)
ZARAGOZA
ARAGÓN
TARRAGONA
Tarragona
Reus
Tortosa
Delta del Ebro
Costa Brava
Costa Dorada
Costa del Azahar
TERUEL
CASTELLÓN (CASTELLÓ)
Castellón (Castelló) de la Plana
Villarreal (Vila-Real)
Sagunto (Sagunt)
VALENCIA
VALENCIA
Torrente (Torrent)
Gandía
Dénia
Golfo de Valencia
Mar Balear
ISLAS BALEARES (ILLES BALEARS)
MAJORCA (MALLORCA)
PALMA
MINORCA (MENORCA)
Mahón (Maó)
Ciudadela (Ciutadella)
IBIZA (EIVISSA)
FORMENTERA
Cabrera
Islas Columbretes
CATALUÑA
ARAGÓN

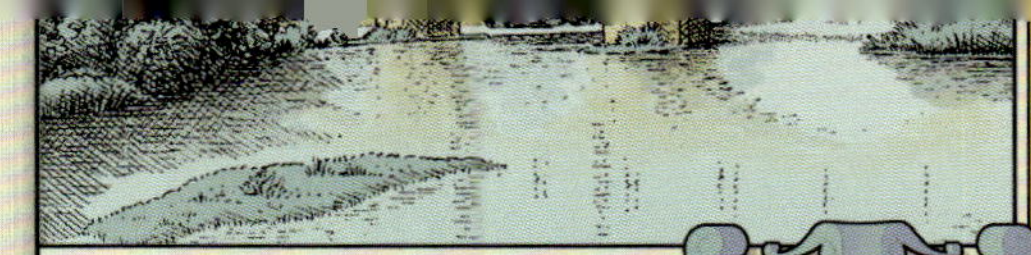

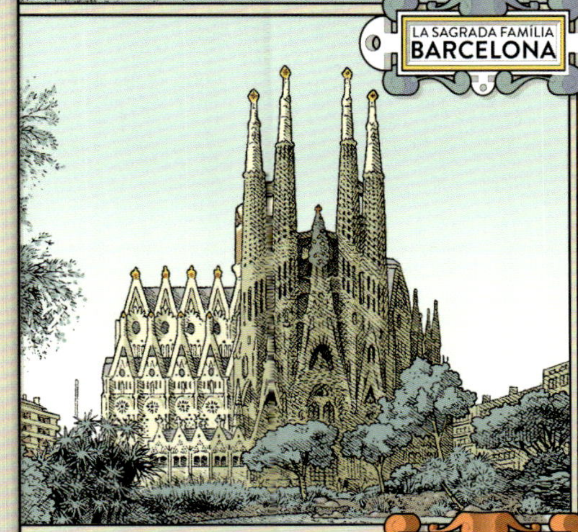

LA SAGRADA FAMILIA
BARCELONA

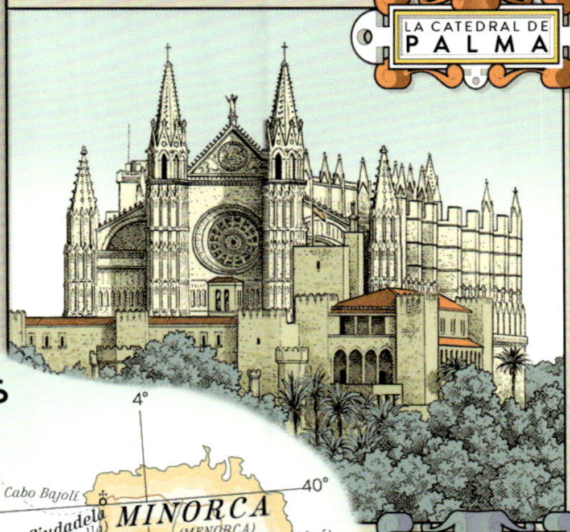

LA CATEDRAL DE
PALMA

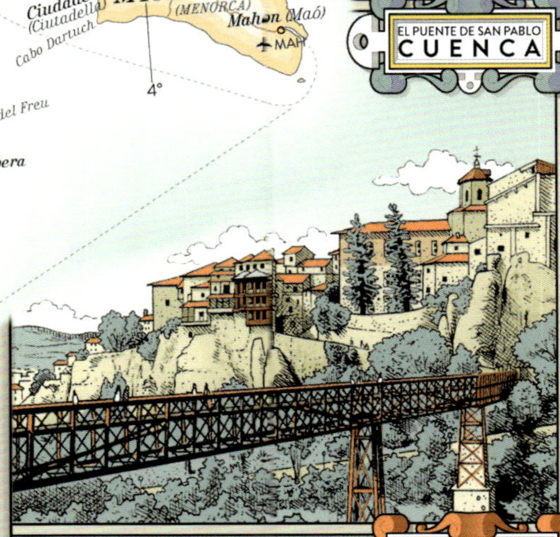

EL PUENTE DE SAN PABLO
CUENCA

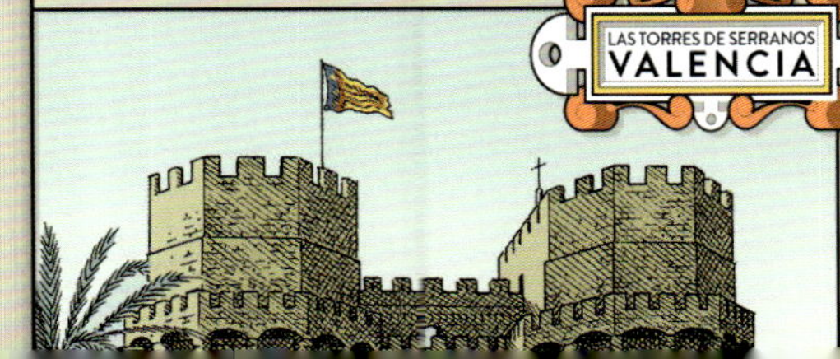

LAS TORRES DE SERRANOS
VALENCIA

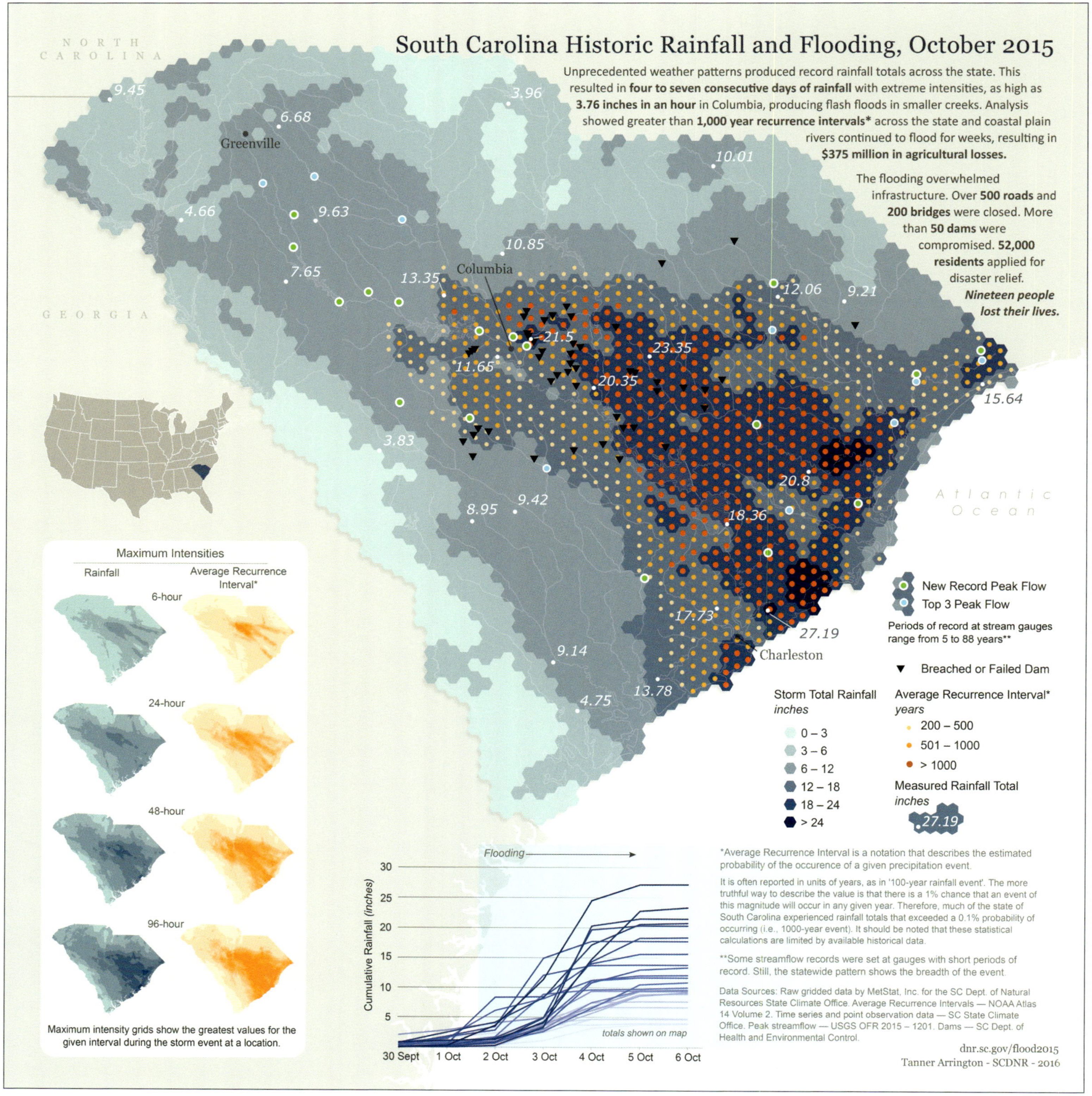
South Carolina Historic Rainfall and Flooding, October 2015
Unprecedented weather patterns produced record rainfall totals across the state. This resulted in four to seven consecutive days of rainfall with extreme intensities, as high as 3.76 inches in an hour in Columbia, producing flash floods in smaller creeks. Analysis showed greater than 1,000 year recurrence intervals* across the state and coastal plain rivers continued to flood for weeks, resulting in $375 million in agricultural losses.
The flooding overwhelmed infrastructure. Over 500 roads and 200 bridges were closed. More than 50 dams were compromised. 52,000 residents applied for disaster relief. Nineteen people lost their lives.
NORTH CAROLINA
GEORGIA
Atlantic Ocean
Greenville
Columbia
Charleston
9.45
6.68
3.96
10.01
4.66
9.63
10.85
7.65
13.35
12.06
9.21
21.5
23.35
11.65
20.35
15.64
3.83
20.8
9.42
8.95
18.36
17.73
27.19
9.14
13.78
4.75
Maximum Intensities
Rainfall
Average Recurrence Interval*
6-hour
24-hour
48-hour
96-hour
Maximum intensity grids show the greatest values for the given interval during the storm event at a location.
Flooding
Cumulative Rainfall (inches)
30
25
20
15
10
5
30 Sept
1 Oct
2 Oct
3 Oct
4 Oct
5 Oct
6 Oct
totals shown on map
New Record Peak Flow
Top 3 Peak Flow
Periods of record at stream gauges range from 5 to 88 years**
Breached or Failed Dam
Storm Total Rainfall
inches
0 – 3
3 – 6
6 – 12
12 – 18
18 – 24
> 24
Average Recurrence Interval*
years
200 – 500
501 – 1000
> 1000
Measured Rainfall Total
inches
27.19
*Average Recurrence Interval is a notation that describes the estimated probability of the occurence of a given precipitation event.
It is often reported in units of years, as in '100-year rainfall event'. The more truthful way to describe the value is that there is a 1% chance that an event of this magnitude will occur in any given year. Therefore, much of the state of South Carolina experienced rainfall totals that exceeded a 0.1% probability of occurring (i.e., 1000-year event). It should be noted that these statistical calculations are limited by available historical data.
**Some streamflow records were set at gauges with short periods of record. Still, the statewide pattern shows the breadth of the event.
Data Sources: Raw gridded data by MetStat, Inc. for the SC Dept. of Natural Resources State Climate Office. Average Recurrence Intervals — NOAA Atlas 14 Volume 2. Time series and point observation data — SC State Climate Office. Peak streamflow — USGS OFR 2015 – 1201. Dams — SC Dept. of Health and Environmental Control.
dnr.sc.gov/flood2015
Tanner Arrington - SCDNR - 2016

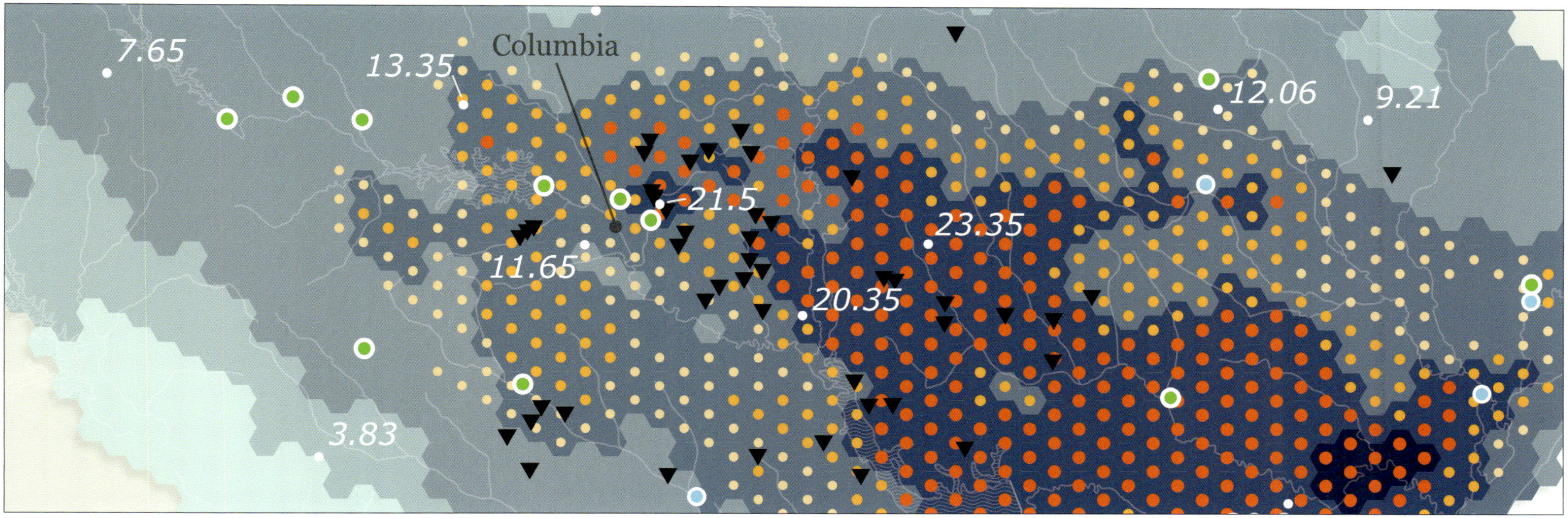

« South Carolina Historic Rainfall and Flooding, October 2015

Tanner Arrington — South Carolina Department of Natural Resources

I woke up early on October 2, 2015, and saw that my rain gauge was nearly full. Columbia, South Carolina, had experienced nearly 10 inches of rain overnight. Significant flash floods were happening across the state, thanks to a unique meteorological situation between Hurricane Joaquin to the south and mid-latitude features to the north. The result was rainfall blasting at us like a giant firehose. The record-setting rainfall continued for days, swelling rivers, overwhelming dams, and washing out roads.

News and social media coverage showed local impacts all over the state: flood waters to the tops of buildings, roads and bridges eroded away, people rescued from their neighborhoods in boats. The images were stirring, but I wanted to add a scientific visualization that would reveal the full breadth and intensity of this meteorological event. Rainfall totals would be most important, but I also wanted to show the rainfall average return interval (ARI), because this event had become known as "the 1,000-year flood."

Generalizing rainfall totals and ARI data into hexagon bins allowed me to show both in the same space. I showed the rainfall values in each hexagon's color, with blue as the obvious choice. I placed the ARI values in the center of each hexagon as an orange dot that varied in size and color. I was pleasantly surprised by the result, which showed patterns in both variables separately and together. I added a few other data points to the map to highlight the storm's other impacts. A simple rainfall timeline and small multiples show the evolution of the event over time.

This map was a team effort. The South Carolina State Climate Office compiled data from many sources to document the event and put it in historical context, providing direction for the map. Guerilla Cartography's *Water: An Atlas* project gave me an opportunity to get peer feedback that improved the final product. And I drew inspiration from the work of countless other cartographers, which helped me visualize this storm event in an interesting and impactful way.

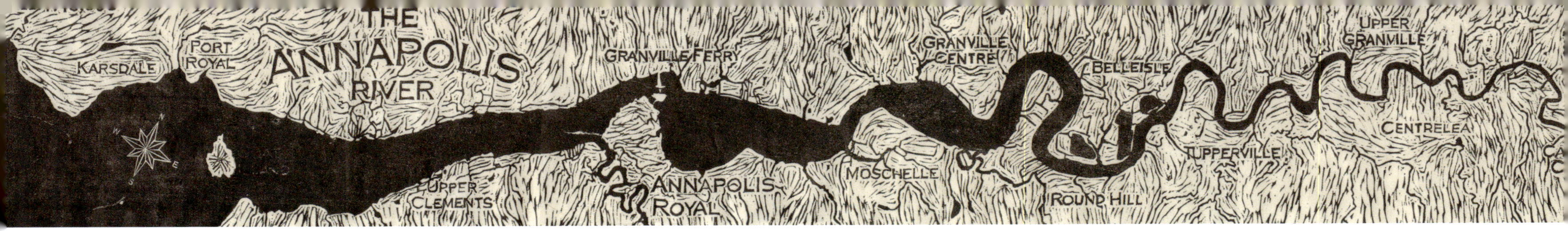

The Annapolis River

Heather Smith

Before the invention of lithography, map design was limited to line work only. The printing processes demanded that each color be printed from a separate plate or added by hand. Intaglio—engraving or etching on metal—was by far the medium of choice for cartography. Almost no one made maps using the cruder, simpler method of woodcut printmaking. It was not practical.

Woodcut prints are made by carving a stamp out of wood, coating it in ink, and pressing it to paper. They were most likely invented before intaglio—around 1400—but there is no period of cartography dominated by woodcuts because the lag time between the two technologies was so brief. Woodcut printing is cheaper than intaglio but does not allow for fine detail. Even when craftsmen reached the highest degree of detail possible, it was still not good enough for cartography. Maps require abundant information—roads, rivers, mountains, and plenty of labels. Good luck carving all that into a block of wood. When woodcut maps were produced, they were either quite simple, such as those in the *Nuremburg Chronicles* (1493) or enormous, such as Jacopo de'Barbari's *View of Venice* (c.1500),

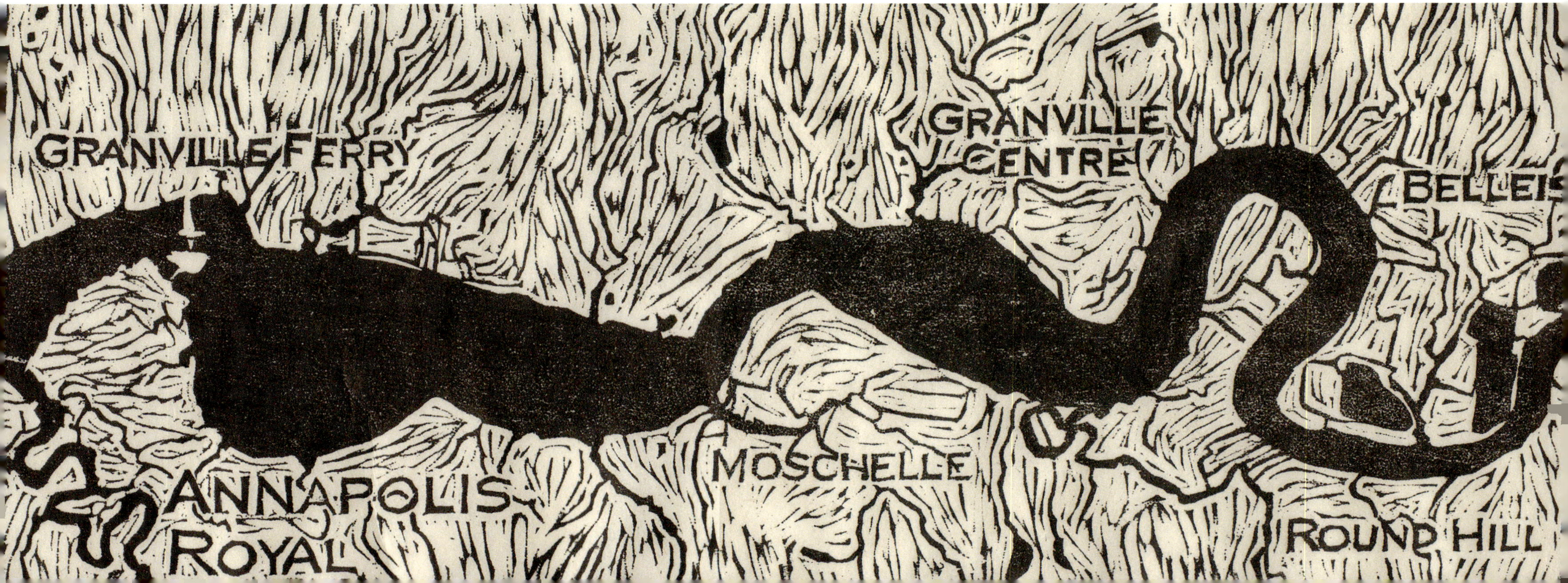

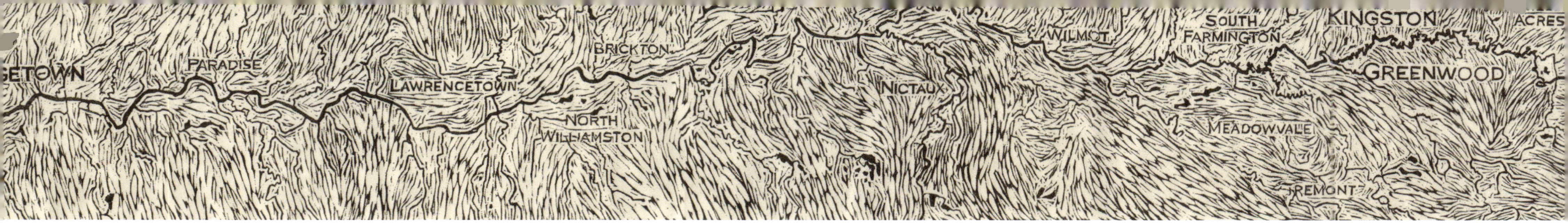

and most were intended to be ephemeral.

This map of the Annapolis River was carved on cherry wood and printed by hand on Japanese paper using oil-based inks. It is six feet long. The texture on the land is mostly fanciful because I allowed the process and material to show through in a piece intended to be more artistic than informative. The text is imperfect, revealing the hand and the time it took to carve.

In making this map, I was limited to simple forms, so I had to choose only the most important features—the river and its towns. The result is a bold graphic image that does not compete with superfluous information. Woodcut is still the most impractical medium for map making. I value it because it forces a quality that is easy to lose in contemporary cartography.

Learn more at heathergabrielsmith.ca

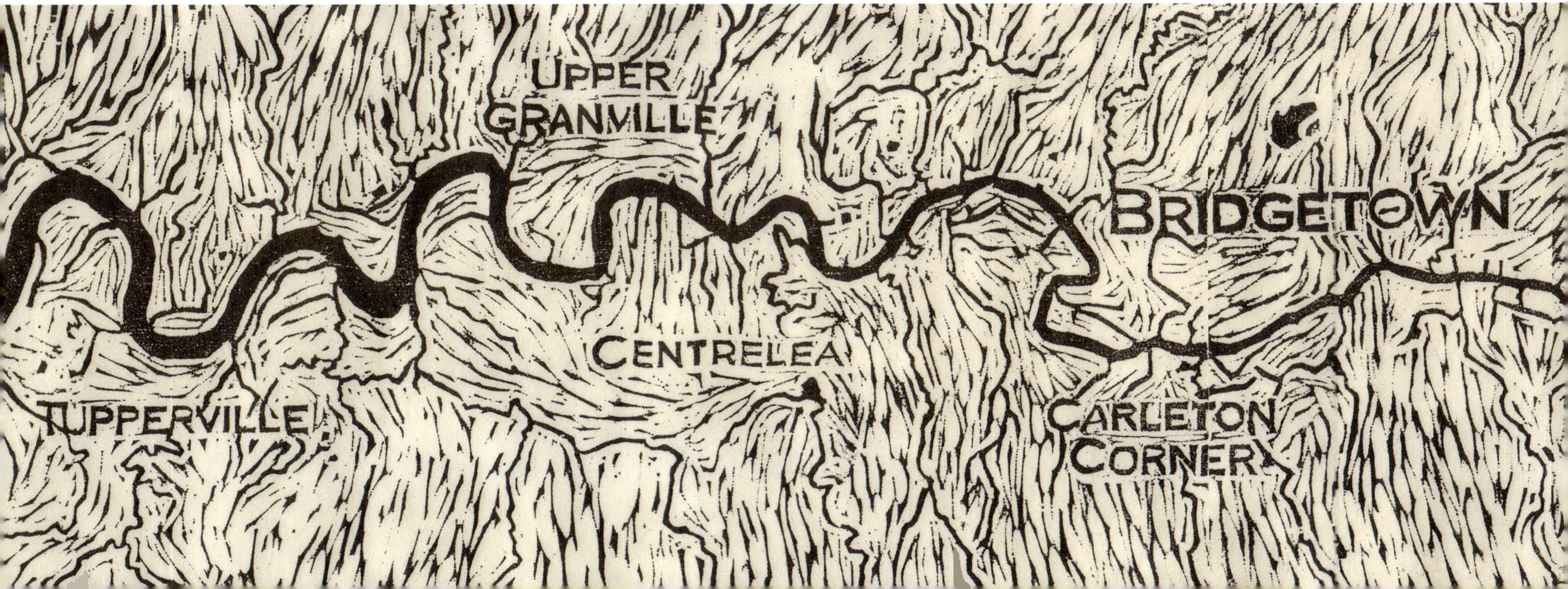

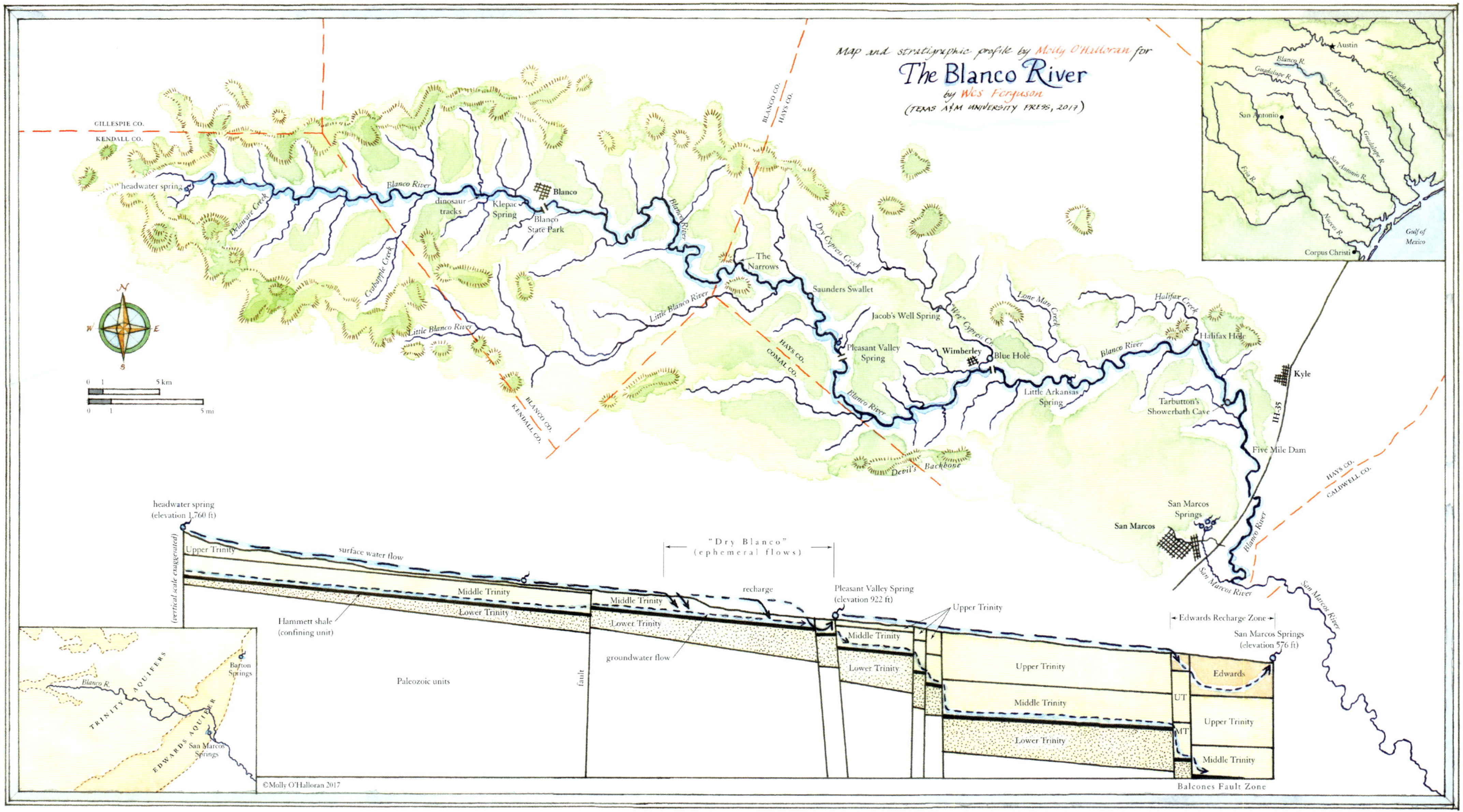
Map and stratigraphic profile by Molly O'Halloran for
The Blanco River
by Wes Ferguson
(Texas A&M University Press, 2017)
Gillespie Co.
Kendall Co.
Blanco Co.
Hays Co.
Comal Co.
Caldwell Co.
headwater spring
Blanco River
Delaware Creek
Crabapple Creek
dinosaur tracks
Klepac Spring
Blanco
Blanco State Park
Little Blanco River
The Narrows
Saunders Swallet
Dry Cypress Creek
Jacob's Well Spring
Pleasant Valley Spring
Wimberley
Blue Hole
Little Arkansas Spring
Lone Man Creek
Halifax Creek
Halifax Hole
Kyle
IH-35
Tarbutton's Showerbath Cave
Five Mile Dam
Devil's Backbone
San Marcos
San Marcos Springs
San Marcos River
0 1 5 km
0 1 5 mi
Austin
Blanco R.
Guadalupe R.
S. Marcos R.
Colorado R.
San Antonio
San Antonio R.
Frio R.
Nueces R.
Corpus Christi
Gulf of Mexico
headwater spring (elevation 1,760 ft)
(vertical scale exaggerated)
surface water flow
"Dry Blanco" (ephemeral flows)
recharge
Pleasant Valley Spring (elevation 922 ft)
Upper Trinity
Middle Trinity
Lower Trinity
Hammett shale (confining unit)
Paleozoic units
fault
groundwater flow
Edwards Recharge Zone
San Marcos Springs (elevation 576 ft)
Edwards
UT
MT
Balcones Fault Zone
Blanco R.
Trinity Aquifers
Edwards Aquifer
Barton Springs
San Marcos Springs
©Molly O'Halloran 2017

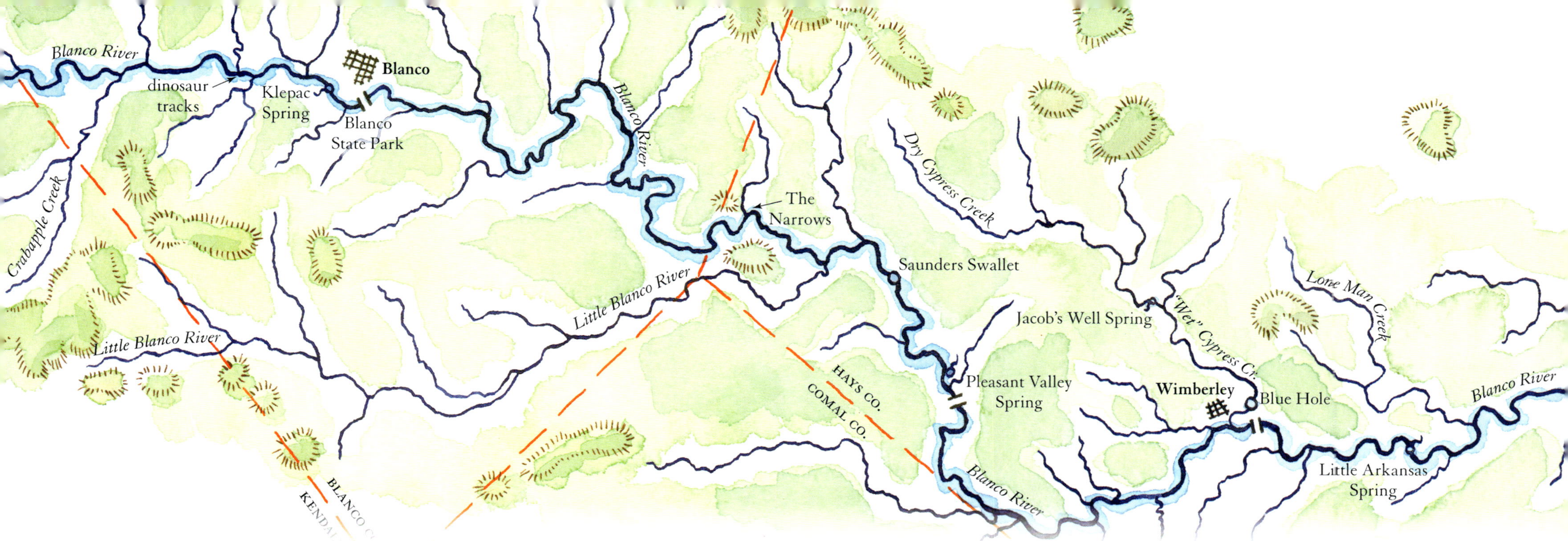

《 The Blanco River

Molly O'Halloran

The Blanco River gets its name from the color of the white limestone bed and cliffs that contain it, but most visitors remember the soft turquoise blue of the water—a refreshing contrast to the dry central Texas landscape. The Blanco—fed by dripping canyon walls, underground springs, and aquifers—is wide and dry for stretches of up to fifteen miles before cutting tall, narrow canyons and bathtub-like swimming holes in the karst. It's a complicated, mysterious river.

This map, created with pen, ink, and watercolor, illustrates the endpapers of Wes Ferguson's book, *The Blanco River*, which is a natural and cultural history featuring color photographs by Jacob Croft Botter. When I begin a map for a book, I always read the introduction, but in this case I kept reading until I'd finished the entire manuscript. Wes's descriptions of exploring the river, discovering its history, and going deeper on field trips with botanists, geologists, and hydrogeologists had me burning through the pages. I wanted to understand exactly how so many physical and cultural features fit together to make the Blanco. And I wanted my map to help readers grasp these connections.

The map's focus, inked and washed in blue, is the twisting, 84-mile length of the Blanco, including all of its tributaries. Hachured hills suggest watershed boundaries while human settlements recede as gray hash marks. An inset map shows the Blanco's waters feeding two more rivers, the San Marcos and Guadalupe, before reaching the Gulf of Mexico. Inspired by an 1836 railroad map, I positioned a geologic cross section directly beneath the plan view so that readers can compare surface features to what's happening underground.

Because most of the Blanco is privately owned and difficult or dangerous to access, we can't walk along the banks of the river to learn its secrets. But by studying the map along with Wes's vivid text, readers can understand the Blanco's complex story above and below the surface.

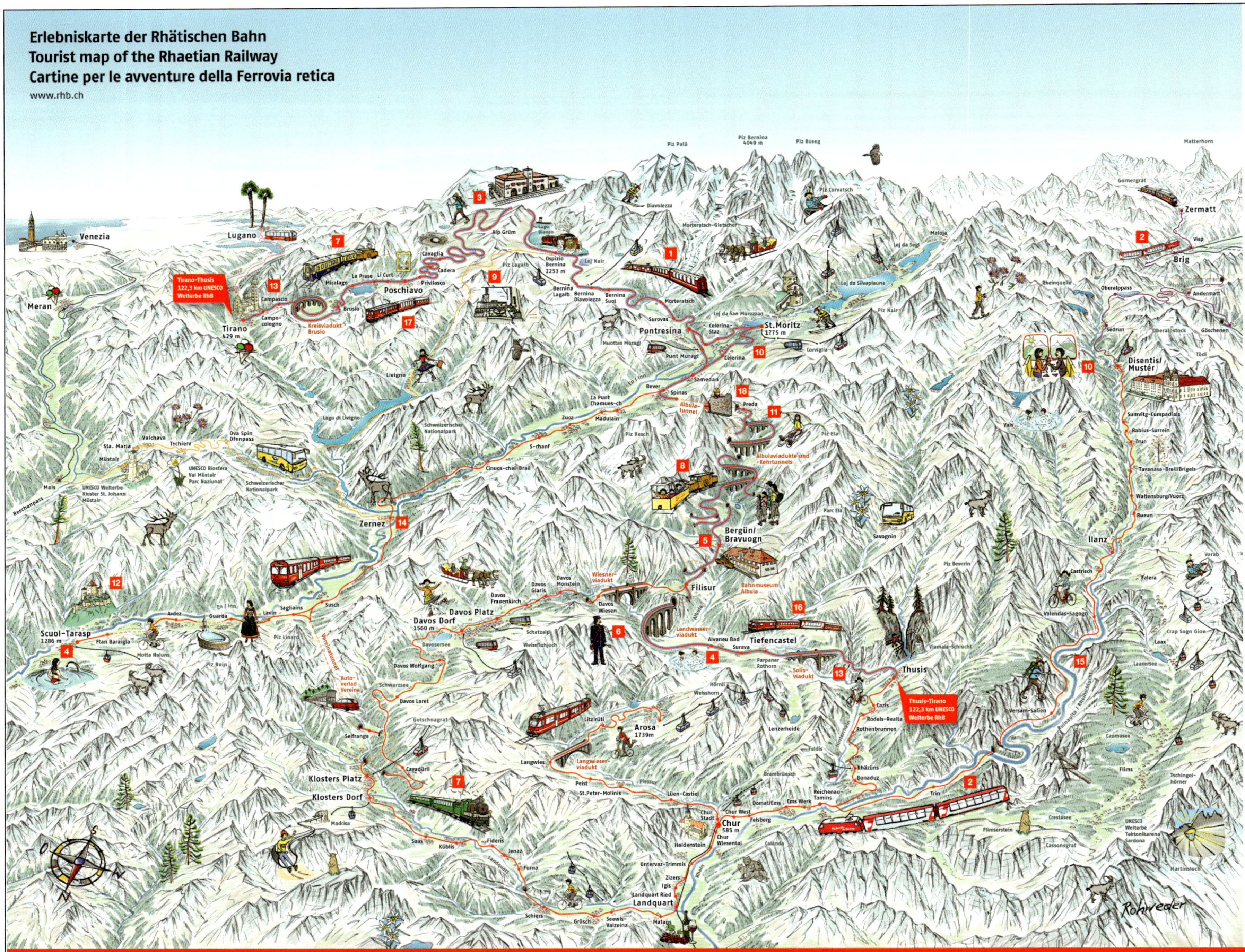

Graubünden

Wie ein roter Faden zieht sich die Rhätische Bahn durch Graubünden, dem grössten Kanton der Schweiz. Reisen Sie mit uns über imposante Brücken, durch tiefe Täler, vorbei an hohen Felsen und wilden Wassern: Erleben Sie magische Momente am laufenden Band!

Graubünden

The Rhaetian Railway wends its way through Graubünden, Switzerland's largest canton like a red thread. Come with us over imposing bridges, through deep valleys and past steep rock faces and wild mountain streams: a succession of magic moments!

Grigioni

La Ferrovia retica si snoda come un filo rosso attraverso i Grigioni, il maggior Cantone della Svizzera. Viaggiate con noi su ponti imponenti, valli profonde, passando vicino ad alte pareti rocciose e acque spumeggianti: godetevi un momento magico dopo l'altro!

Die Rhätische Bahn in Zahlen
The Rhaetian Railway in figures
La Ferrovia retica in cifre

Eröffnung erste Strecke
First track opened | Inaugurazione della prima tratta: 1889 (Landquart – Davos)

Streckenlänge | Track length | Lunghezza complessiva: 384 000 m

Spurweite | Gauge | Scartamento: 1000 mm

Mitarbeiter | Employees | Collaboratori: 1500

Anzahl Brücken | Number of bridges | Numero di ponti: 606

Längste Brücke | Longest bridge | Ponte più lungo: 285 m (Langwieserviadukt | Langwieser Viaduct | viadotto di Langwiesen)

Höchste Brücke | Highest bridge | Ponte più alto: 89 m (Wiesnerviadukt | Wiesner Viaduct | viadotto di Wiesner)

Anzahl Tunnels | Number of tunnels | Numero di gallerie: 115

Grösste Steigung | Steepest gradient | Pendenza massima: 70 ‰ (Bernina Linie | Bernina Line | linea del Bernina)

Höchster Punkt | Highest point | Punto più elevato: 2 253 m ü. M. | m a.s.l. | m s.l.m. (Ospizio Bernina)

Tiefster Punkt | Lowest point | Punto più basso: 429 m ü. M. | m a.s.l. | m s.l.m. (Tirano)

Personenbeförderungen | Passenger volume
Persone trasportate: 10 000 000 jährlich | per year | all'annc

Die Höhepunkte auf dem RhB-Streckennetz
The highlights of the RhB rail network
Gli highlights della rete FR

1. **Bernina Express** www.berninaexpress.ch
2. **Glacier Express** www.glacierexpress.ch
3. **Alp Grüm** www.rhb.ch/berninalinie
4. **Bade-Kombi Engadin Bad Scuol / Bad Alvaneu**
   Spa package | Treno e terme
   www.rhb.ch/badscuol / www.rhb.ch/badalvaneu
5. **Bahnmuseum Albula**
   Albula Railway Museum | Museo ferroviario dell'Albula
   www.rhb.ch/bahnmuseum
6. **Clà Ferrovia der Kinder-Kondukteur**
   Children's conductor | Il controllore per i bambini
   www.rhb.ch/cla
7. **Dampf- & Nostalgiefahrten**
   Steam- & nostalgic rides | Viaggi su treni a vapore e nostalgici
   www.rhb.ch/dampffahrten
8. **Erlebniszug Albula**
   Albula Adventure Train | Treno avventura dell'Albula
   www.rhb.ch/erlebniszug
9. **Führerstandsfahrten**
   Footplate rides | Macchinisti per un giorno
   www.rhb.ch/fuehrerstandsfahrten
10. **Kulinarische Genussreisen**
    Culinary trips | Viaggi gastronomici
    www.rhb.ch/kulinarik
11. **Schlittelwelt Preda/Darlux – Bergün**
    Sledging | Il paradiso dello slittino
    www.rhb.ch/schlittelwelt
12. **Schloss Tarasp**
    The castle of Tarasp | Castello di Tarasp
    www.rhb.ch/schlosstarasp
13. **Via Albula/Bernina** www.rhb.ch/via
14. **Nationalparkzentrum Zernez**
    National Park Centre Zernez | Centro del Parco Nazionale Zernez
    www.rhb.ch/nationalpark
15. **Rheinschlucht/Ruinaulta**
    Rhine Gorge/Ruinaulta | Gola del Reno/Ruinaulta
    www.rhb.ch/ruinaulta
16. **Albula Panorama Winter**
    Albula Panorama Winter | Albula Panorama Inverno
    www.rhb.ch/albulapanorama
17. **Bernina Panorama Winter**
    Bernina Panorama Winter | Bernina Panorama Inverno
    www.rhb.ch/berninapanorama
18. **Infoarena Albulatunnel**
    Albula Tunnel Infoarena | Infoarena del tunnel dell'Albula
    www.rhb.ch/infoarena

Rhätische Bahn

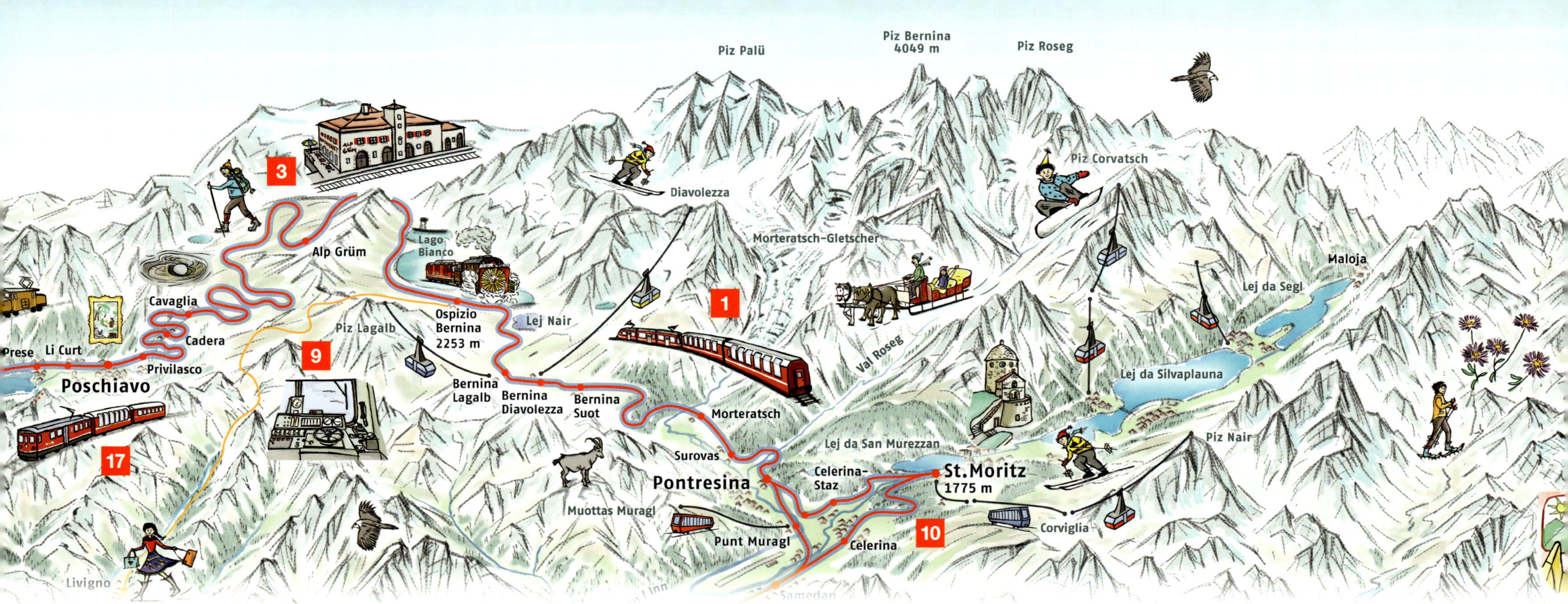

« Rhaetian Railway Map

Arne and Martina Rohweder

The Albula and Bernina lines of the Rhaetian Railway in the Swiss Alps are two historic railways that move passengers over major mountain passes. The routes pass through 55 tunnels and covered galleries and cross 196 viaducts and bridges in their combined 128-kilometer journeys. The Albula and Bernina lines have made a major impact on the cultures and economies of otherwise isolated mountain communities and are exceptional examples of technical, environmental, and architectural design.

The Albula and Bernina routes were added to the UNESCO World Heritage List in 2008. We produced a special map for marketing this unique railway line from 2012 to 2017. We did the work in cooperation with Süsskind Graphic Design and Communication, which developed the concept of a playful panoramic map. The style of painting recalls a famous children's book, *A Bell for Ursli*, drawn by local illustrator Alois Carigiet.

Arne Rohweder sketched the landscape with a chalk pen and painted the shading with a watercolor pen—first in subtle summer colors and then in grayish-blue tones for the winter version. The map includes different perspectives that give the viewer glimpses into otherwise hidden valleys and gorges. A view from Grisons to faraway Valais and on to the Matterhorn shows the railway line on which the carriages of the Glacier Express run to Zermatt. Martina Rohweder illustrated the many tourist attractions. The vintage look of the illustrations reminds viewers of the period around 1940, but with the addition of contemporary recreational activities.

We hope to encourage the audience to engage in local sports, excursions, and leisure travel, so the map is updated frequently, and the illustrations highlight a variety of activities in the region. The map's popularity has exceeded our expectations and now has a cult following!

Learn more at panoramakarte.com

FRENTE DE LUTA POR MODADIA

FLM

Proliferating in São Paulo's vacant fissures and crevices, an occupied city is germinating. As in numerous derelict buildings and sites around the world, homeless movements are piecing together self-made shelters. This is the temporary city of São Paulo's FLM, the 'Front of the Housing Struggle' is at the vanguard of this planetary city in the making. Its map will be outdated before being finished, as they will be evicted in one place, only to squat again elsewhere.

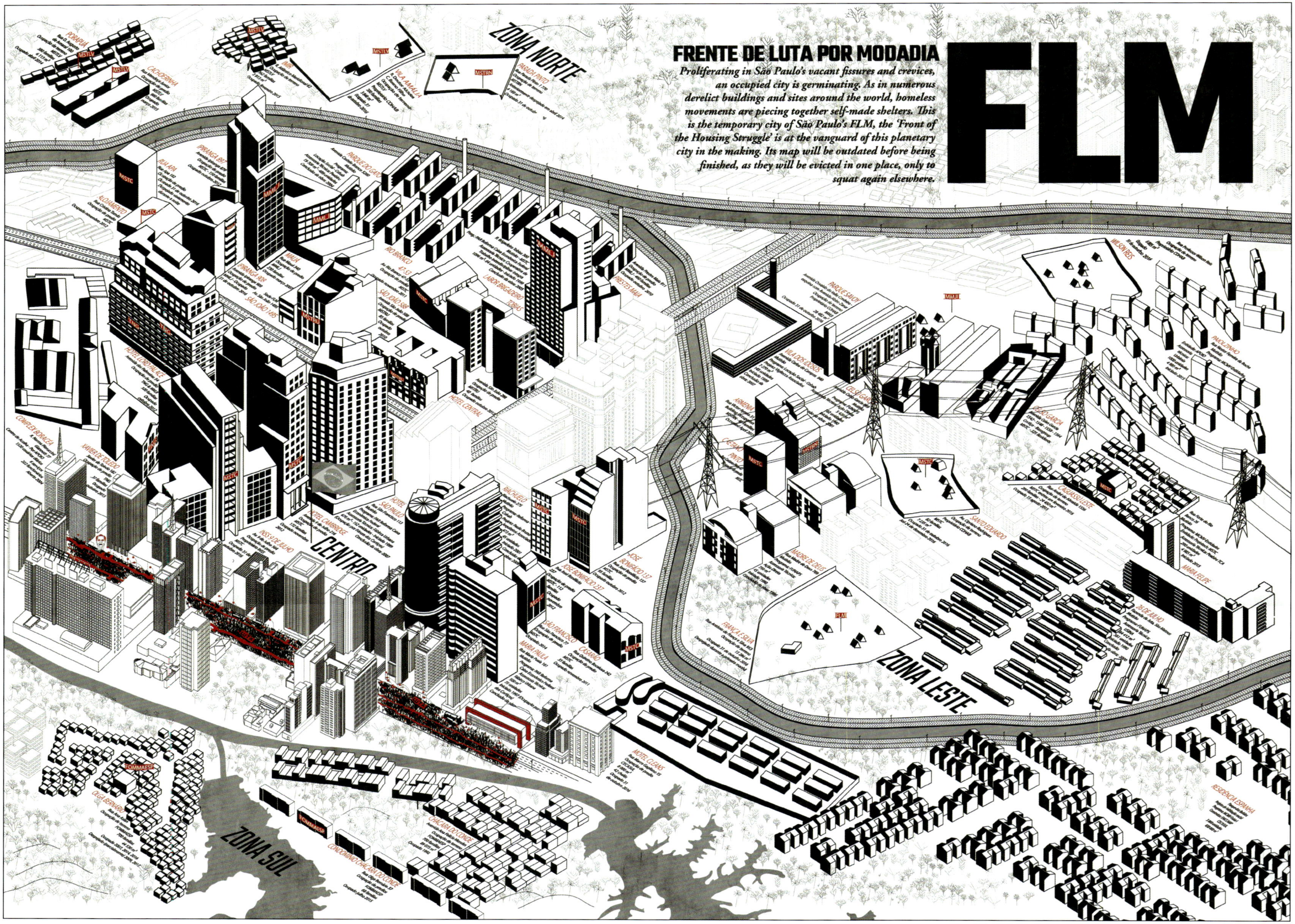

Occupied Central São Paulo »

« Frente de Luta por Moradia (FLM)

Jeroen Stevens, Ken Vervaet, Matthias Lamberts, Lucas Lerchs, Kathleen De Beukelaer, Claire Bosmans, Valentine Van Den Eynde Raissa Monteiro, Carmen Briers, Lisa De Vos, and Bruno De Meulder

Dispersed over São Paulo's urban sprawl is a city-within-a-city unlike any other. *The Frente de Luta por Moradia*, one of São Paulo's most powerful social movements, is piecing together a vast occupied territory, perpetually in the making and relentlessly on the move. This coalition of homeless movements accommodates thousands of people in squatter settlements, vacant buildings, and conquered apartments. Some occupations persist for a few hours; others endure for decades. As soon as a map of the occupied city is drawn, it is outdated, so this map captures only a representative snapshot of São Paulo as it continuously transforms. The map sketches out a still moment in perpetual urban motion.

The isometric FLM map singles out hidden spaces in the megalopolis of São Paulo that were occupied by the housing movement at the beginning of 2017, providing a glimpse of a city that is usually invisible. The occupied city is partly handmade, recycling formerly vacant spaces. But the movement also capitalizes on existing remarkable urban structures, including large avenues, transport systems, and iconic modernist buildings. On the map, canals and principal avenues provide basic orientation. Thin line drawings identify essential landscape context, including iconic buildings, water reservoirs, and patches of preserved forest. Bold black solids mark currently occupied buildings and terrains, situated in the city center and on its northern, eastern, and southern periphery. Red signs on the map reproduce the FLM's self-representation, marking its presence with flags, banners, and tags across the city.

The occupied city is permanently temporary, never providing a stable home; it is also temporarily permanent, never really disappearing. The map on the right represents the spread of vacant, underused or misused, and occupied locations in white, grey and red solid fills. The timeline on the left vertically charts occupied buildings from 1970 to 2017, representing their endurance on the horizontal axis. The chart exposes how people move from one building to another, evicted in one place and squatting again elsewhere. The diagram explains how buildings are captured in a circular process of use, underuse, abandonment, and occupation. Each part of the drawing functions as a legend for the other, using simplified white-grey-black coding to unravel the interplay between occupations on the one hand, and cyclical processes of urban development on the other.

This cartographic exploration stems from years of collaborative work with homeless movements in São Paulo. The maps seek to contribute to their politics of *presence*, underscoring their agency in carving out spaces of shared living for the lowest income populations in the metropolis.

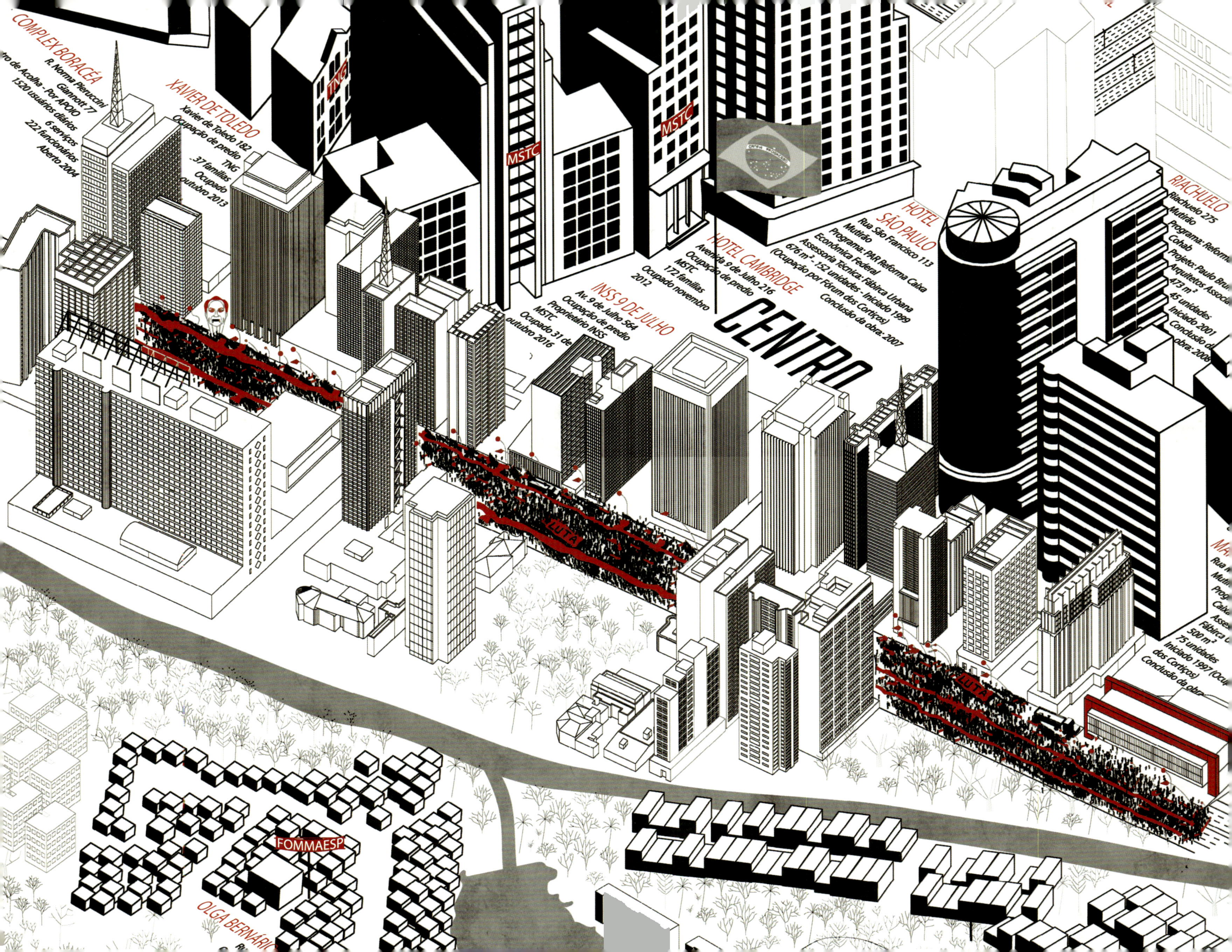

COMPLEX BORACÉA
R. Norma Pieruccini
Giannott 77
Centro de Acolha - Por APOIO
1520 usuários diários
6 serviços
222 funcionários
Aberto 2004
XAVIER DE TOLEDO
Xavier de Toledo 182
Ocupação de predio
TNG
37 famílias
Ocupado
outubro 2013
TNG
MSTC
MSTC
HOTEL
SÃO PAULO
Rua São Francisco 113
Mutirão
Programa: PAR Reforma Caixa
Econômica Federal
Assessoria Técnica: Fábirca Urbana
676 m² - 152 unidades - Iniciado 1999
(Ocupação por Fórum dos Cortiços)
Conclusão da obra: 2007
HOTEL CAMBRIDGE
Avenida 9 de Julho 216
Ocupação de predio
MSTC
172 famílias
Ocupado novembro
2012
INSS 9 DE JULHO
Av. 9 de Julho 564
Ocupação de predio
Proprietário INSS
MSTC
Ocupado 31 de
outubro 2016
CENTRO
RIACHUELO
Riachuelo 275
Mutirão
Programa: Reforma
Cohab
Projeto: Paulo Bru
Arquitetos Associa
473 m²
45 unidades
Iniciado 2001
Conclusão d
obra: 2008
LUTA
LUTA
LUTA
500 m²
75 unidades
Iniciado 1997 (Ocu
dos Cortiços)
Conclusão da obra
FOMMAESP
OLGA BERNÁRIO

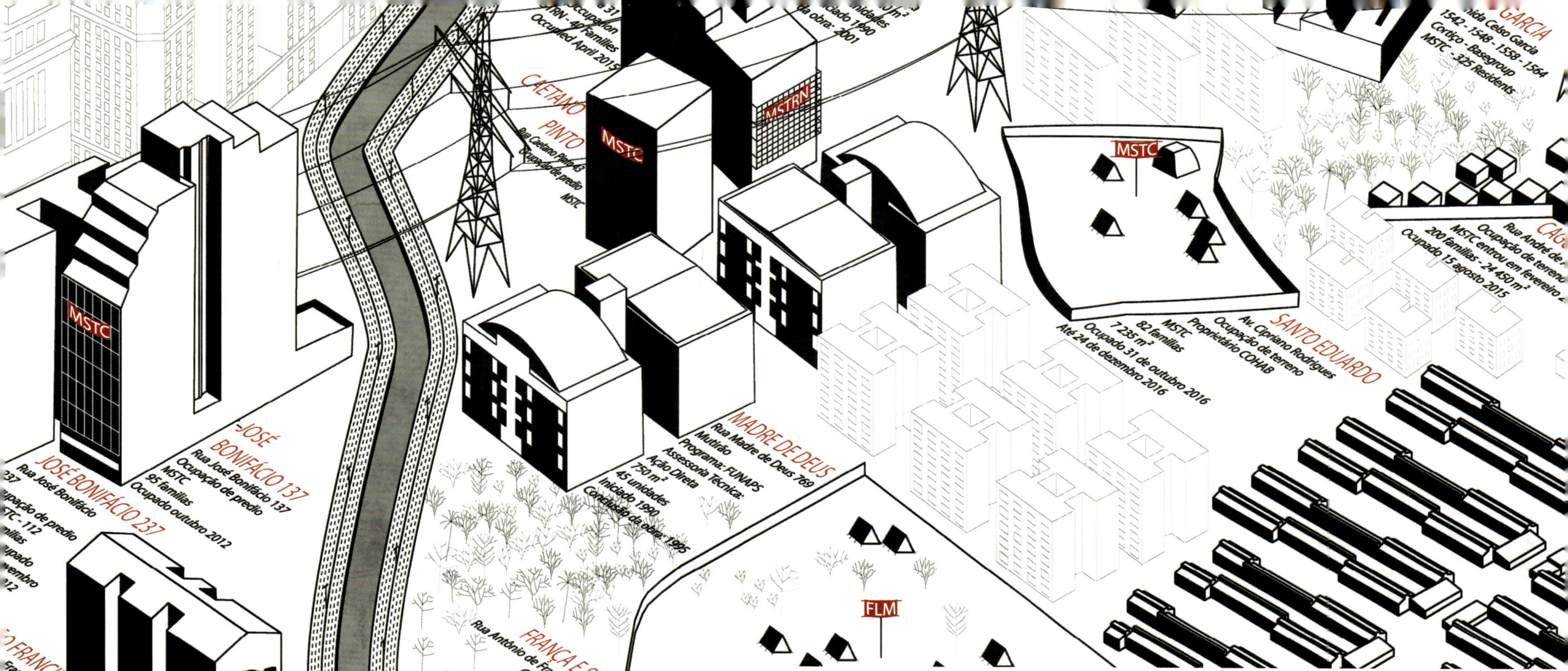
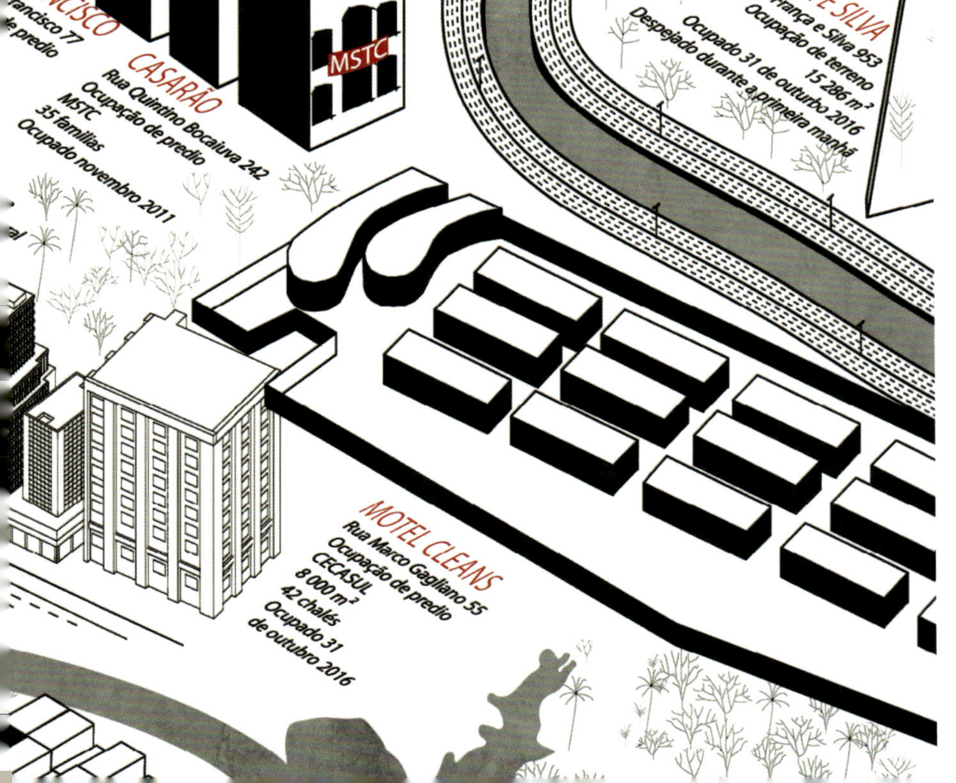

CAETANO PINTO
Rua Caetano Pinto 45
Ocupação de predio
MSTC
MSTC
MSTRN
GARCIA
1542 - 1548 - 1558 - 1564
Cortiço - Basegroup
MSTC - 325 Residents
MSTC
SANTO EDUARDO
Av. Cipriano Rodrigues
Ocupação de terreno
Proprietário COHAB
MSTC
82 famílias
7 235 m²
Ocupado 31 de outubro 2016
Até 24 de dezembro 2016
Ocupação de terreno
200 famílias - 24 450 m²
Ocupado 15 agosto 2015
MSTC
MADRE DE DEUS
Rua Madre de Deus 769
Mutirão
Programa: FUNAPS
Assessoria Técnica:
Ação Direta
750 m²
45 unidades
Iniciado 1990
Conclusão da obra: 1995
JOSÉ BONIFÁCIO 137
Rua José Bonifácio 137
Ocupação de predio
MSTC
95 famílias
Ocupado outubro 2012
JOSÉ BONIFÁCIO 237
Rua José Bonifácio
FRANÇA E SILVA
Rua Antônio de França e Silva 953
Ocupação de terreno
15 286 m²
Ocupado 31 de outubro 2016
Despejado durante a primeira manhã
FLM
MSTC
CASARÃO
Rua Quintino Bocaiuva 242
Ocupação de predio
MSTC
35 famílias
Ocupado novembro 2011
MOTEL CLEANS
Rua Marco Gagliano 55
Ocupação de predio
CECASUL
8 000 m²
42 chalés
Ocupado 31
de outubro 2016

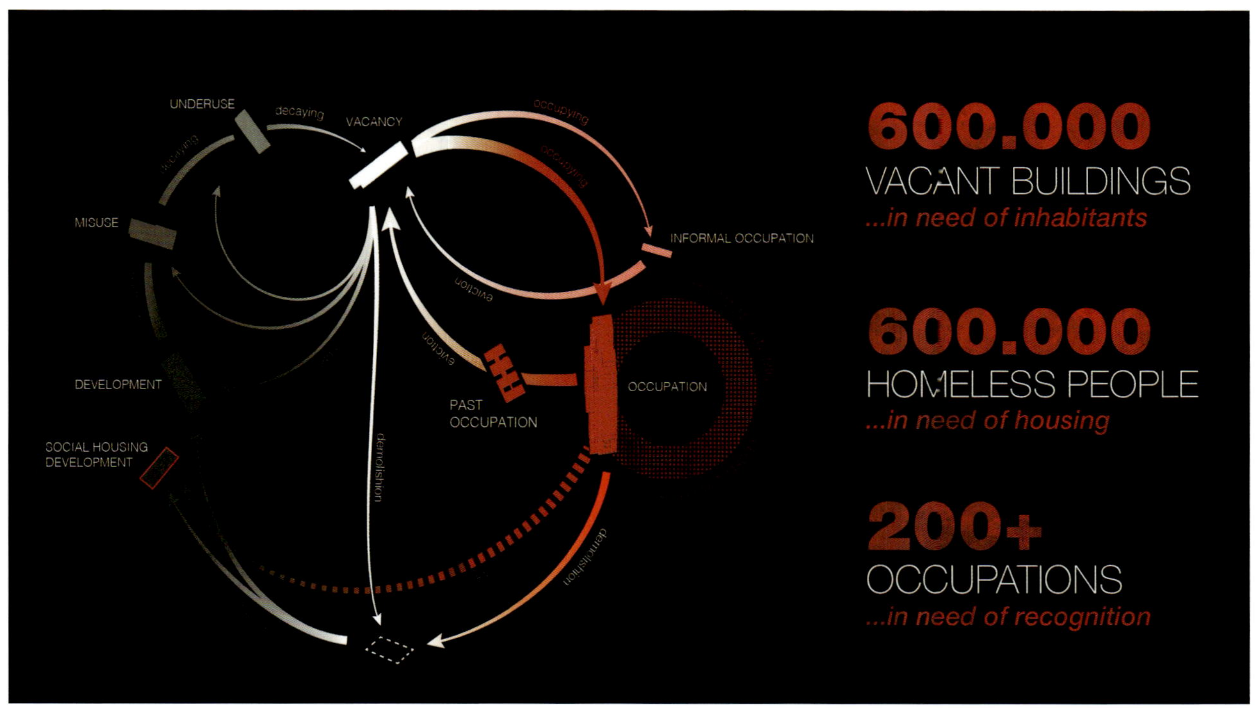

UNDERUSE
decaying
VACANCY
occupying
MISUSE
INFORMAL OCCUPATION
eviction
DEVELOPMENT
PAST OCCUPATION
OCCUPATION
SOCIAL HOUSING DEVELOPMENT
demolishment
600.000
VACANT BUILDINGS
...in need of inhabitants
600.000
HOMELESS PEOPLE
...in need of housing
200+
OCCUPATIONS
...in need of recognition

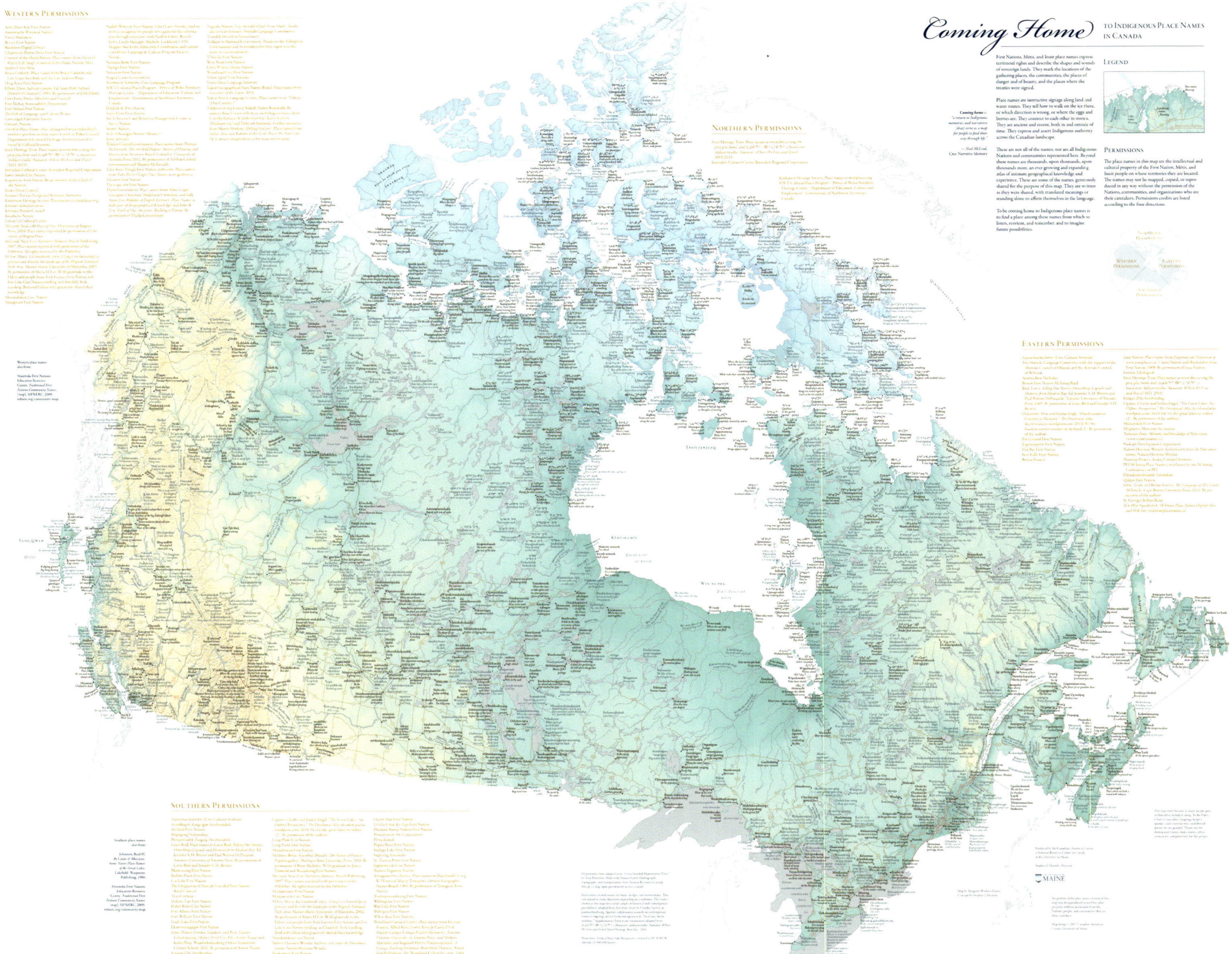
Coming Home
to Indigenous Place Names in Canada
First Nations, Métis, and Inuit place names express territorial rights and describe the shapes and sounds of sovereign lands. They mark the locations of the gathering places, the communities, the places of danger and of beauty, and the places where the treaties were signed.
Place names are instructive signage along land and water routes. They tell how to walk on the ice there, or which direction is wrong, or where the eggs and berries are. They connect to each other in stories. They are ancient and recent, both in and outside of time. They express and assert Indigenous authority across the Canadian landscape.
These are not all of the names, nor are all Indigenous Nations and communities represented here. Beyond these names are thousands, upon thousands, upon thousands more, an ever growing and expanding atlas of intimate geographical knowledge and experience. These are some of the names, generously shared for the purpose of this map. They are written as they were shared, with translated meanings or standing alone to affirm themselves in the language.
To be coming home to Indigenous place names is to find a place among these names from which to listen, reorient, and remember, and to imagine future possibilities.
Coming home is "a return to Indigenous memories and narratives [that] serve as a map for people to find their way through life."
— Neal McLeod, Cree Narrative Memory
Legend
Permissions
The place names in this map are the intellectual and cultural property of the First Nation, Métis, and Inuit people on whose territories they are located. The names may not be mapped, copied, or reproduced in any way without the permission of the Nations, communities, and organizations who are their caretakers. Permissions credits are listed according to the four directions.
Western Permissions
Northern Permissions
Eastern Permissions
Southern Permissions

《 Coming Home to Indigenous Place Names in Canada

Margaret Wickens Pearce, for the Canadian-American Center, University of Maine

Sometimes I feel like throwing rocks,
but I would be afraid
that I would lose my name
and lose my territory
and damage my Mother.
- ssipsis, 1972

To honor the names and their caretakers.
To protect sovereignty.
To foreground Indigenous protocols.
To get it right, for once.

To make something beautiful.
To contribute something useful.
To specify.
To provoke questions.

To be quiet and listen.

To structure projection, size, and fold in dialogue with maps by Natural Resources Canada.
To represent borders with place names, as they are experienced.
To indicate that place names participate at all time scales.
To let meaning come from an accumulation of feeling.

To be wary of my assumptions as they continue to come up.
To be okay with not knowing what's next.
To remember it's none of my business.
To trust the design process to show the way.

To work against complacency, condescension, and dystopia.
To shift societal expectations for what a map can do or be.
To call it cartography, not counter-cartography.
To remind people that print is not broken.

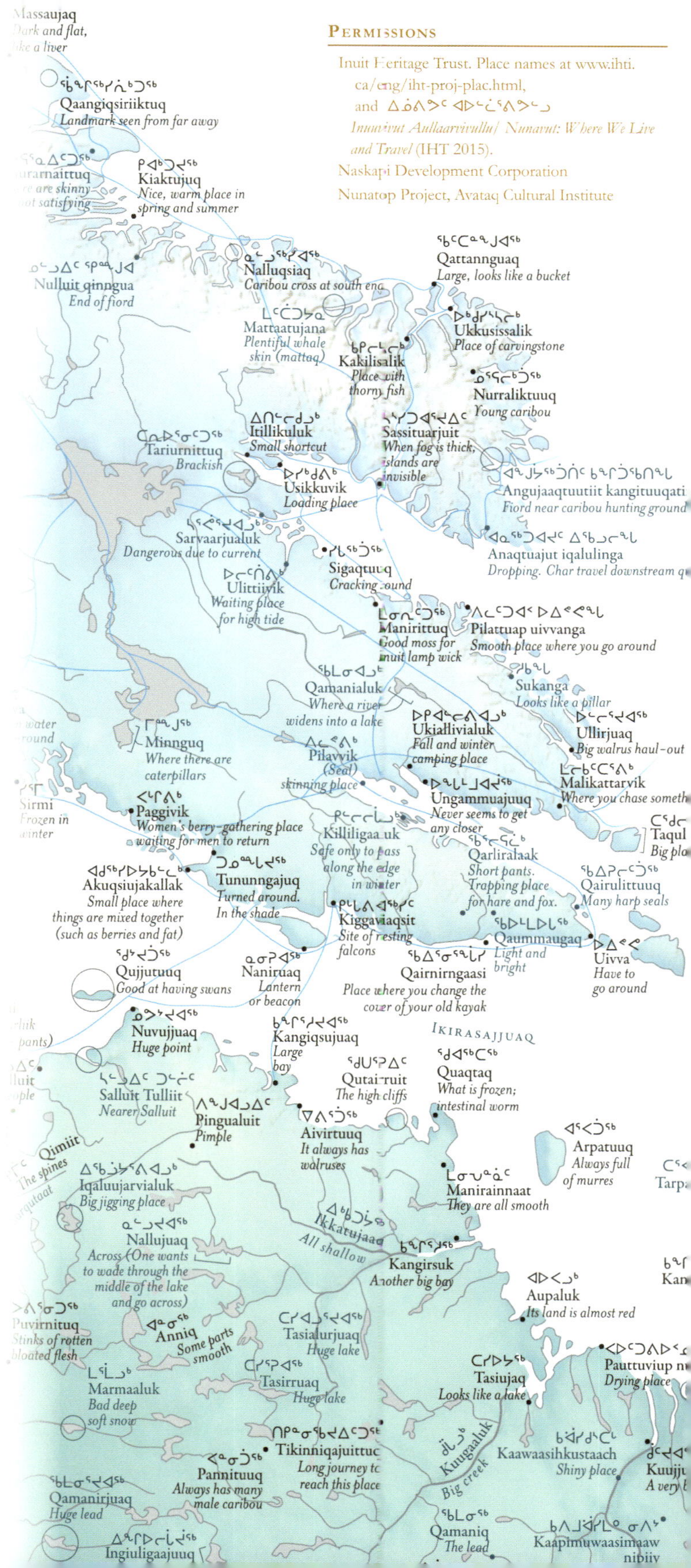

Permissions

Inuit Heritage Trust. Place names at www.ihti.ca/eng/iht-proj-plac.html, and ᐃᓅᐱᐅᑉ ᐊᐅᓚᐅᕐᕕᐅᓪᓗ *Inuuvirut Aullaarvivullu/ Nunavut: Where We Live and Travel* (IHT 2015).
Naskapi Development Corporation
Nunatop Project, Avataq Cultural Institute

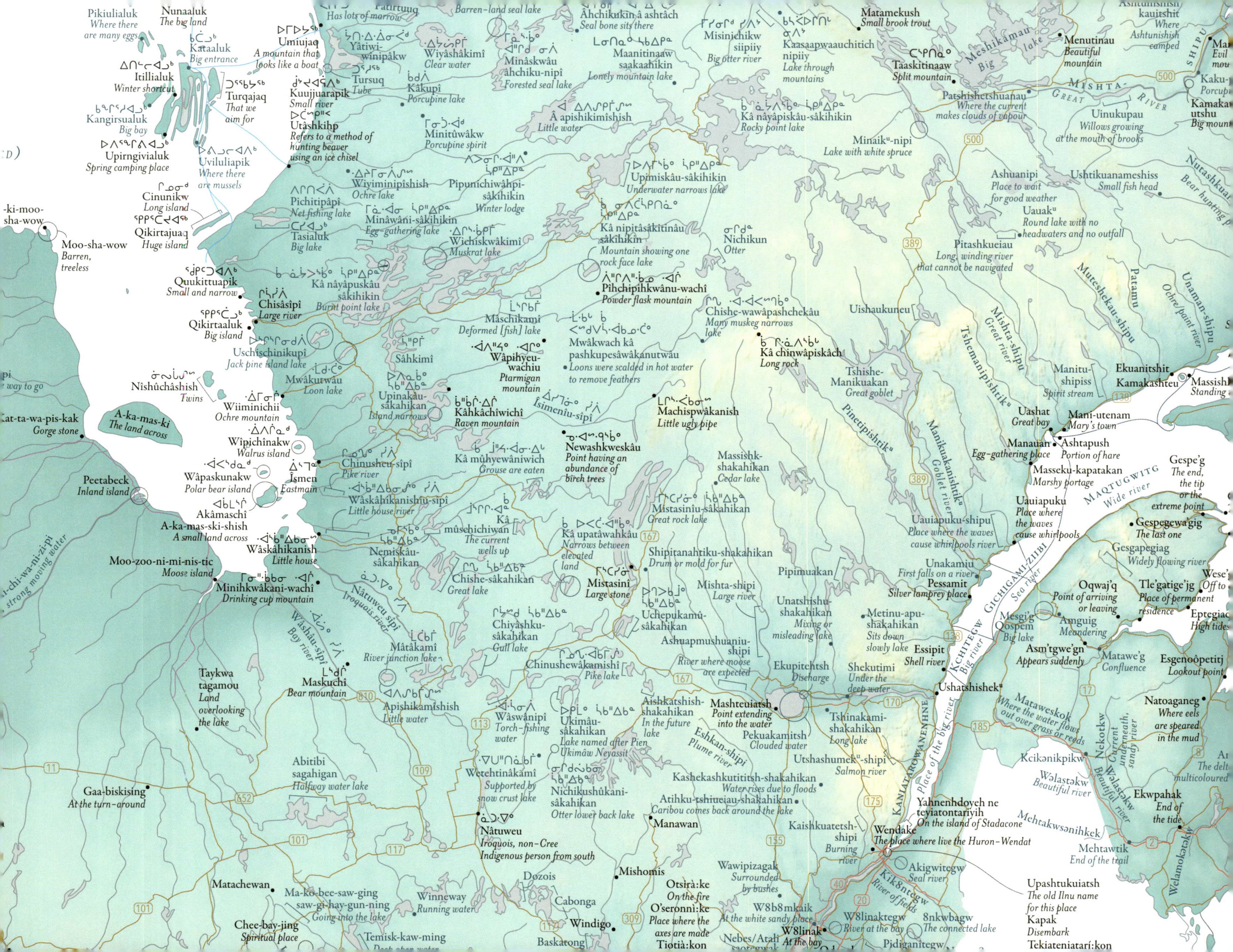

Pikiulialuk
Where there are many eggs
Nunaaluk
The big land
Kataaluk
Big entrance
Umiujaq
A mountain that looks like a boat
Itillialuk
Winter shortcut
Turqajaq
That we aim for
Kangirsualuk
Big bay
Upirngivialuk
Spring camping place
Uviluliapik
Where there are mussels
Cinunikw
Long island
Qikirtajuaq
Huge island
-ki-moo-sha-wow
Moo-sha-wow
Barren, treeless
Quukittuapik
Small and narrow
Qikirtaaluk
Big island
Nîshûchâshish
Twins
Kat-ta-wa-pis-kak
Gorge stone
A-ka-mas-ki
The land across
Wiiminichiii
Ochre mountain
Wîpichînakw
Walrus island
Wâpaskunakw
Polar bear island
Peetabeck
Inland island
Akâmaschî
A-ka-mas-ki-shish
A small land across
Wâskâhikanish
Little house
Moo-zoo-ni-mi-nis-tic
Moose island
Minihkwâkani-wachî
Drinking cup mountain
strong moving water
Taykwa tagamou
Land overlooking the lake
Maskuchî
Bear mountain
Abitibi sagahigan
Halfway water lake
Gaa-biskising
At the turn-around
Matachewan
Ma-ko-bee-saw-ging saw-gi-hay-gun-ning
Going into the lake
Chee-bay-jing
Spiritual place
Temisk-kaw-ming
Has lots of marrow
Yâtiwi-winipâkw
Tursuq
Tube
Kuujjuarapik
Small river
Utâshkihp
Refers to a method of hunting beaver using an ice chisel
Wiyâshâkimî
Clear water
Kâkupî
Porcupine lake
Minâskwâu âhchiku-nipî
Forested seal lake
Barren-land seal lake
Âhchikukin â ashtâch
Seal bone sits there
Maanitinaaw saakaahikin
Lonely mountain lake
Â apishikimishish
Little water
Minitûwâkw
Porcupine spirit
Wiyiminipîshish
Ochre lake
Pipunichiwâhpi-sâkihikin
Winter lodge
Pichitipâpî
Net fishing lake
Minâwâni-sâkihikin
Egg-gathering lake
Tasialuk
Big lake
Wichiskwâkimî
Muskrat lake
Kâ nâyâpuskâu sâkihikin
Burnt point lake
Chisâsîpî
Large river
Uschischinikupî
Jack pine island lake
Mwâkurwâu
Loon lake
Sâhkimî
Mâschikamî
Deformed [fish] lake
Wâpihyeu-wachiu
Ptarmigan mountain
Upinakâu-sâkahikan
Island narrows
Kâhkâchîwichî
Raven mountain
Isimeniu-sîpî
Chinusheu-sîpî
Pike river
Îsmen
Eastmain
Kâ mûhyewâniwich
Grouse are eaten
Wâskâhîkanishîu-sîpî
Little house river
Kâ mûsehichiwan
The current wells up
Nemiskâu-sâkahikan
Chishe-sâkahikan
Great lake
Nâtuweu sîpî
Iroquois river
Wâshâu-sîpî
Bay river
Mâtâkamî
River junction lake
Chiyâshku-sâkahikan
Gull lake
Apishikamîshish
Little water
Wâswânipî
Torch-fishing water
Wetehtinâkamî
Supported by snow crust lake
Nâtuweu
Iroquois, non-Cree Indigenous person from south
Winneway
Running water
Dozois
Cabonga
Windigo
Baskatong
Upimiskâu-sâkihikin
Underwater narrows lake
Kâ nipitâsâkitinâu sâkihikin
Mountain showing one rock face lake
Pihchipihkwânu-wachî
Powder flask mountain
Mwâkwach kâ pashkupesâwâkanutwâu
Loons were scalded in hot water to remove feathers
Newashkweskâu
Point having an abundance of birch trees
Kâ upatâwahkâu
Narrows between elevated land
Mistasinî
Large stone
Uchepukamû-sâkahikan
Chinushewâkamishî
Pike lake
Ukimâu-sâkahikan
Lake named after Pien Ukimâw Neyassit
Nichikushûkani-sâkahikan
Otter lower back lake
Mishomis
Misinichikw siipiiy
Big otter river
Kaasaapwaauchitich nipiiy
Lake through mountains
Kâ nâyâpiskâu-sâkihikin
Rocky point lake
Nichikun
Otter
Chishe-wawâpashchekâu
Many muskeg narrows lake
Kâ chinwâpiskâch
Long rock
Machispwâkanish
Little ugly pipe
Massishk-shakahikan
Cedar lake
Mistasiniu-sâkahikan
Great rock lake
Shipitanahtiku-shakahikan
Drum or mold for fur
Mishta-shipi
Large river
Ashuapmushuaniu-shipi
River where moose are expected
Aishkatshish-shakahikan
In the future lake
Eshkan-shipi
Plume river
Mashteuiatsh
Point extending into the water
Pekuakamitsh
Clouded water
Kashekashkutititsh-shakahikan
Water rises due to floods
Atihku-tshiueiau-shakahikan
Caribou comes back around the lake
Manawan
Otsirà:ke
On the fire
O'seronni:ke
Place where the axes are made
Tiotià:kon
Wawipizagak
Surrounded by bushes
W8b8mkaik
At the white sandy place
Nebes/Atali
W8linak
At the bay
Matamekush
Small brook trout
Taaskitinaaw
Split mountain
Minaiku-nipi
Lake with white spruce
Patshishetshuanau
Where the current makes clouds of vapour
Meshikamau
Big lake
Menutinau
Beautiful mountain
Mishta-shipu
Great river
Uinukupau
Willows growing at the mouth of brooks
Ashtunishish
Where Ashtunishish camped
Shipu
Evil
Kaku-
Porcup
Kamaka utshu
Big mount
Ashuanipi
Place to wait for good weather
Ushtikuanameshiss
Small fish head
Nutashkuan
Bear hunting
Uauaku
Round lake with no headwaters and no outfall
Pitashkueiau
Long, winding river that cannot be navigated
Uishaukuneu
Tshemanipishtiku
Mishta-shipu
Great river
Muteshekau-shipu
Patamu
Unaman-shipu
Ochre point river
Tshishe-Manikuakan
Great goblet
Pinetipishtiku
Manikuakanishtiku
Goblet river
Manitu-shipiss
Spirit stream
Ekuanitshit
Kamakashteu
Massish
Standing
Uashat
Great bay
Mani-utenam
Mary's town
Manauan
Egg-gathering place
Ashtapush
Portion of hare
Masseku-kapatakan
Marshy portage
Uauiapuku
Place where the waves cause whirlpools
Uauiapuku-shipu
Place where the waves cause whirlpools river
Unakamiu
First falls on a river
Pessamit
Silver lamprey place
Pipimuakan
Unatshishu-shakahikan
Mixing or misleading lake
Metinu-apu-shakahikan
Sits down slowly lake
Essipit
Shell river
Ekupitchtsh
Discharge
Shekutimi
Under the deep water
Ushatshishek
Tshinakami-shakahikan
Long lake
Utshashumeku-shipi
Salmon river
Kaishkuatetsh-shipi
Burning river
Wendake
The place where live the Huron-Wendat
Yahnenhdoyeh ne teyiatontariyih
On the island of Stadacone
Akigwitegw
Seal river
Kik8ntegw
River of fields
W8linaktegw
River at the bay
8nkwbagw
The connected lake
Pidiganitegw
Kaniatarowanenhne
Place of the big river
Kchitegw
Big river
Gichigami-ziibi
Sea river
Gespe'g
The end, the tip or the extreme point
Maqtugwitg
Wide river
Gespegewa'gig
The last one
Gesgapegiag
Widely flowing river
Wese'
Off to
Oqwaj'q
Point of arriving or leaving
Tle'gatige'jg
Place of permanent residence
Eptegiac
High tides
Mesgi'g Qospem
Big lake
Amguig
Meandering
Asm'tgwe'gn
Appears suddenly
Matawe'g
Confluence
Esgenoôpetitj
Lookout point
Mataweskok
Where the water flows out over grass or reeds
Natoaganeg
Where eels are speared in the mud
Nekotkw
Current underneath sandy river
Kcikənikpikw
Wəlastəkw
Beautiful river
Walastakw
Beautiful river
Ekwpahak
End of the tide
Mehtakwsənihkek
Mehtawtik
End of the trail
Welamokotakw
The delta multicoloured
Upashtukuiatsh
The old Ilnu name for this place
Kapak
Disembark
Tekiateniatarí:kon

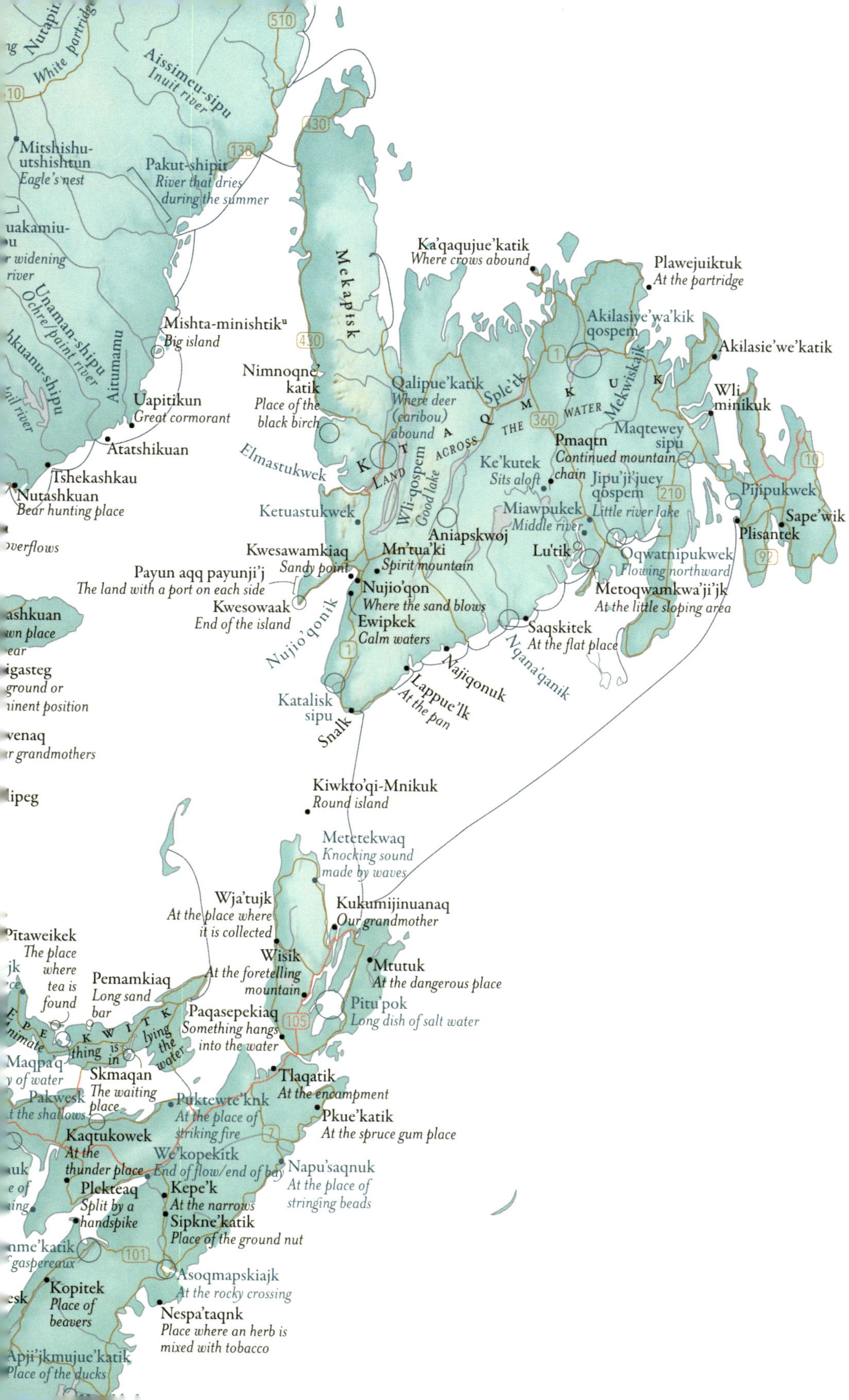

Permissions

Aanischaaukamikw (Cree Cultural Institute)
The Abenaki Language Committee with the support of the Abenaki Council of Odanak and the Abenaki Council of Wôlinak
Andrea Bear Nicholas
Benoit First Nation Mi'kmaq Band
Bird, Louis. *Telling Our Stories: Omushkego Legends and Histories from Hudson Bay.* Ed. Jennifer S. H. Brown and Paul Warren DePasquale. Toronto: University of Toronto Press, 2005. By permission of Louis Bird and Jennifer S.H. Brown.
Delaronde, Hiio and Jordan Engle. "Haudenosaunee Country in Mohawk." *The Decolonial Atlas* decolonialatlas.wordpress.com/2015/02/04/haudenosaunee-country-in-mohawk-2/. By permission of the authors.
Eel Ground First Nation
Esgenoôpetitj First Nation
Flat Bay First Nation
Fort Folly First Nation
Bernie Francis
Fort Albany First Nation
Innu Nation. Place names from *Pepamuteiati Nitassinat* at www.innuplaces.ca. © Innu Nation and Sheshatshiu Innu First Nation, 2008. By permission of Innu Nation.
Institut Tshakapesh
Inuit Heritage Trust. Place names at www.ihti.ca/eng/iht-proj-plac.html, and ᐃᓅᕕᕗᑦ ᐊᐅᓪᓛᕐᕕᕗᓪᓗ *Inuuvivut Aullaarvivullu/ Nunavut: Where We Live and Travel* (IHT 2015).
Kitigan Zibi Anishinabeg
Lippert, Charles and Jordan Engel. "The Great Lakes: An Ojibwe Perspective." *The Decolonial Atlas* decolonialatlas.wordpress.com/2015/04/14/the-great-lakes-in-ojibwe-v2/. By permission of the authors.
Long Point First Nation
Miawpukek First Nation
Mi'gmawei Mawiomi Secretariat
Nametau Innu: Memory and knowledge of Nitassinan (www.nametauinnu.ca)
Naskapi Development Corporation
Nation Huronne-Wendat Archives et Centre de Documentation, Nation Huronne-Wendat
Nunatop Project, Avataq Cultural Institute
PEI Mi'kmaq Place Names contributed by the Mi'kmaq Confederacy of PEI
Pekuakamiulnuatsh Takuhikan
Qalipu First Nation
Sable, Trudy and Bernie Francis. *The Language of This Land, Mi'kma'ki.* Cape Breton University Press, 2012. By permission of the authors.
St. George's Indian Band
Ta'n Weji-Squalia'tiek: Mi'kmaw Place Names Digital Atlas and Web Site (mikmawplacenames.ca)
Taykwa Tagamou Nation
Temagami First Nation. Place names in Macdonald, Craig K. *Historical Map of Temagami.* Ontario Geographic Names Board, 1985. By permission of Temagami First Nation.

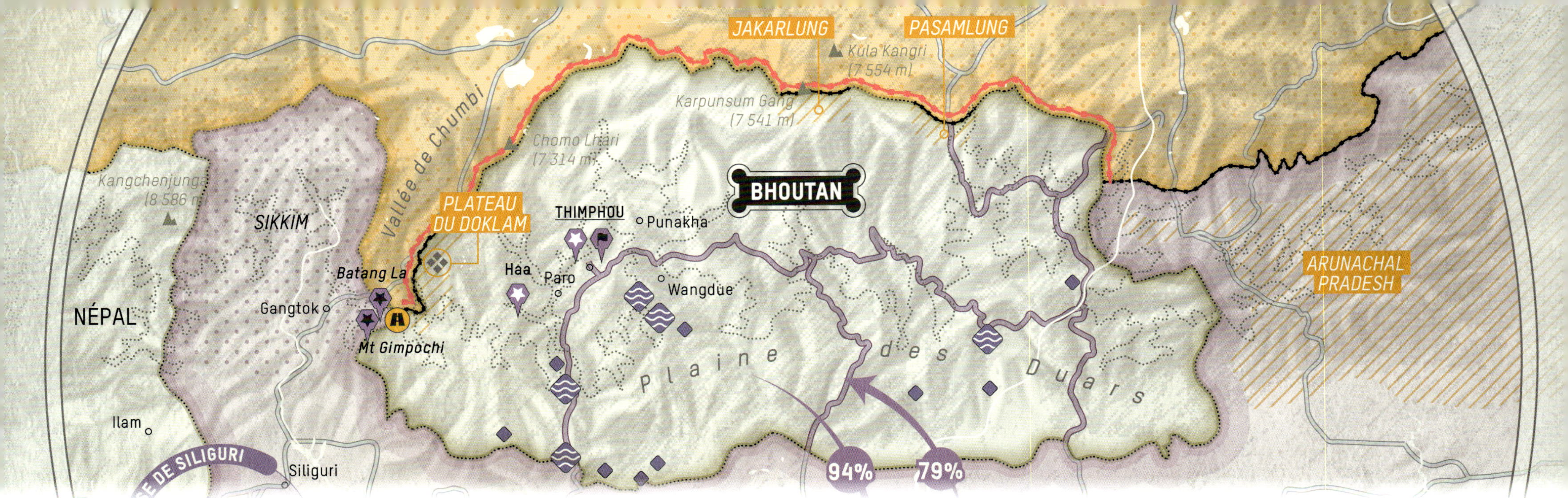

Bhoutan, Un Confetti Himalayen Lorgné par Deux Géants, » Bhutan, a Himalayan "Confetti" Ogled by Two Giants

Francesca Fattori and Xemartin Laborde — Le Monde

This cartographic portrait of Bhutan appeared first in the French newspaper *Le Monde*. In a weekly feature, the paper devotes a full page to a story—in maps and infographics—of a particular place and issue. We have great journalistic freedom in selecting the topic and aesthetic treatment. We are fortunate to be able to explore a different region each week with unique graphic representations.

The infographic is divided into several parts: a central map, surrounded by contextual information in legends; a timeline highlighting major dates in the history of Bhutan; and a regional map of South Asia. The dominant geometrical shape is the circle, which has the advantage of focusing the viewer's attention on the focal point of the page—the map—and highlighting additional information in a harmonious way.

Duotone—the key graphic and pedagogical choice—carries through all of the elements on the page. The "two giants" mentioned in the title are China, in orange, and India, in purple. These tart but sweet colors help us represent the Buddhist universe, honoring the dominant religion in Bhutan. The reader can understand the country's geopolitical situation visually, as it is surrounded by the giants. To highlight significant geopolitical tensions, red and black represent diplomatic breaks and disputed borders.

Finally, we added elevation to the image. Shaded relief and a curved dotted line showing the 3,000-meter elevation mark illustrate the physical context of Bhutan, perched high in the Himalayan peaks. We also included a radiation effect and a photo texture that bring depth to the page. The legend titles mirror the orange and purple colors used in the maps to make a strong visual connection.

The goal of this particular work was to bring attention to Bhutan's isolation and dependence on its neighbors using all our cartographic and artistic tools.

Un confetti himalayen lorgné par deux géants

Le Bhoutan, petit pays pauvre coincé entre la Chine et l'Inde, doit composer avec ses puissants voisins

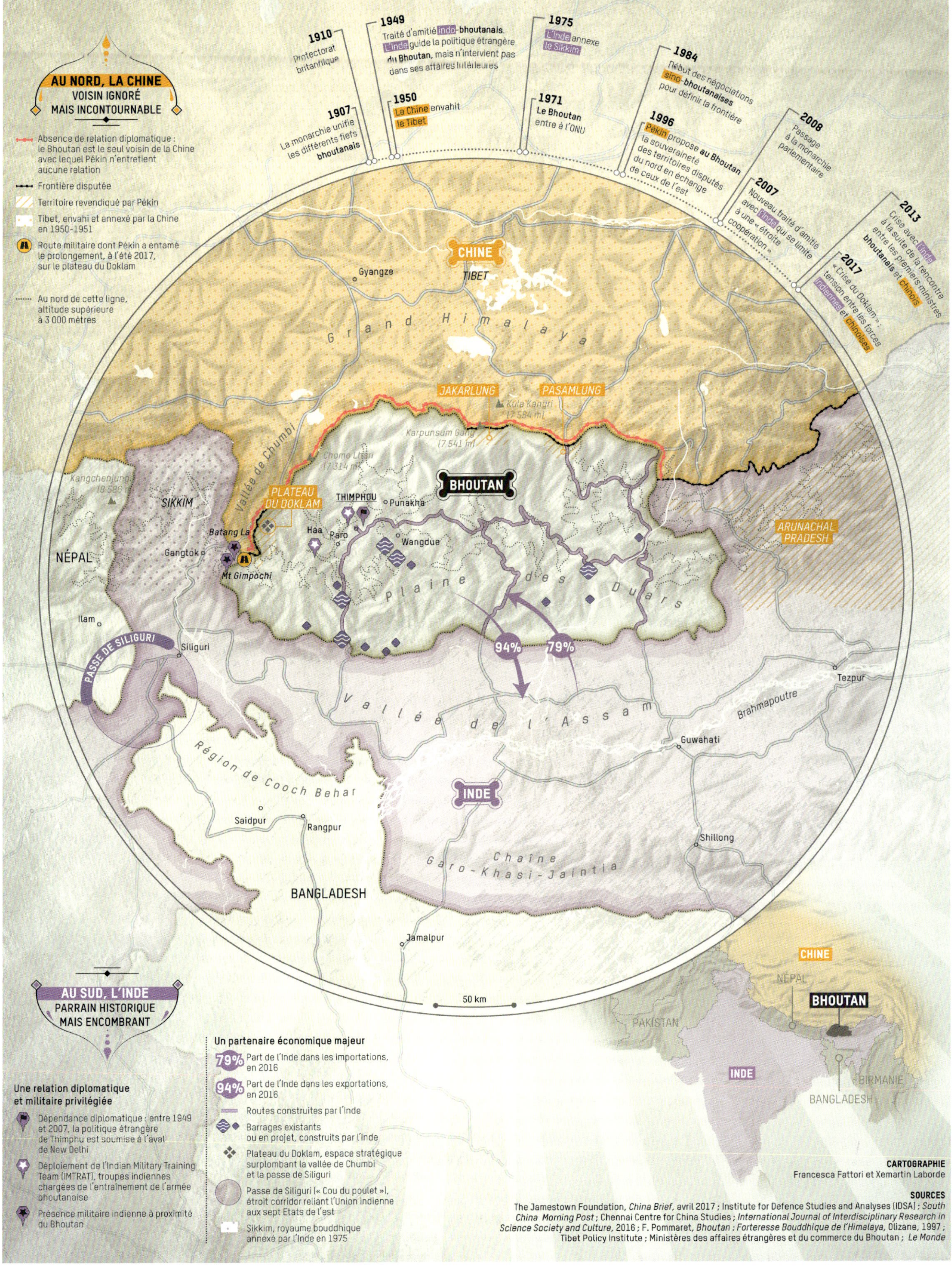

CARTOGRAPHIE
Francesca Fattori et Xemartin Laborde

SOURCES
The Jamestown Foundation, *China Brief*, avril 2017 ; Institute for Defence Studies and Analyses (IDSA) ; *South China Morning Post* ; Chennai Centre for China Studies ; *International Journal of Interdisciplinary Research in Science Society and Culture*, 2016 ; F. Pommaret, *Bhoutan : Forteresse Bouddhique de l'Himalaya*, Olizane, 1997 ; Tibet Policy Institute ; Ministères des affaires étrangères et du commerce du Bhoutan ; *Le Monde*

Katmai Nationa

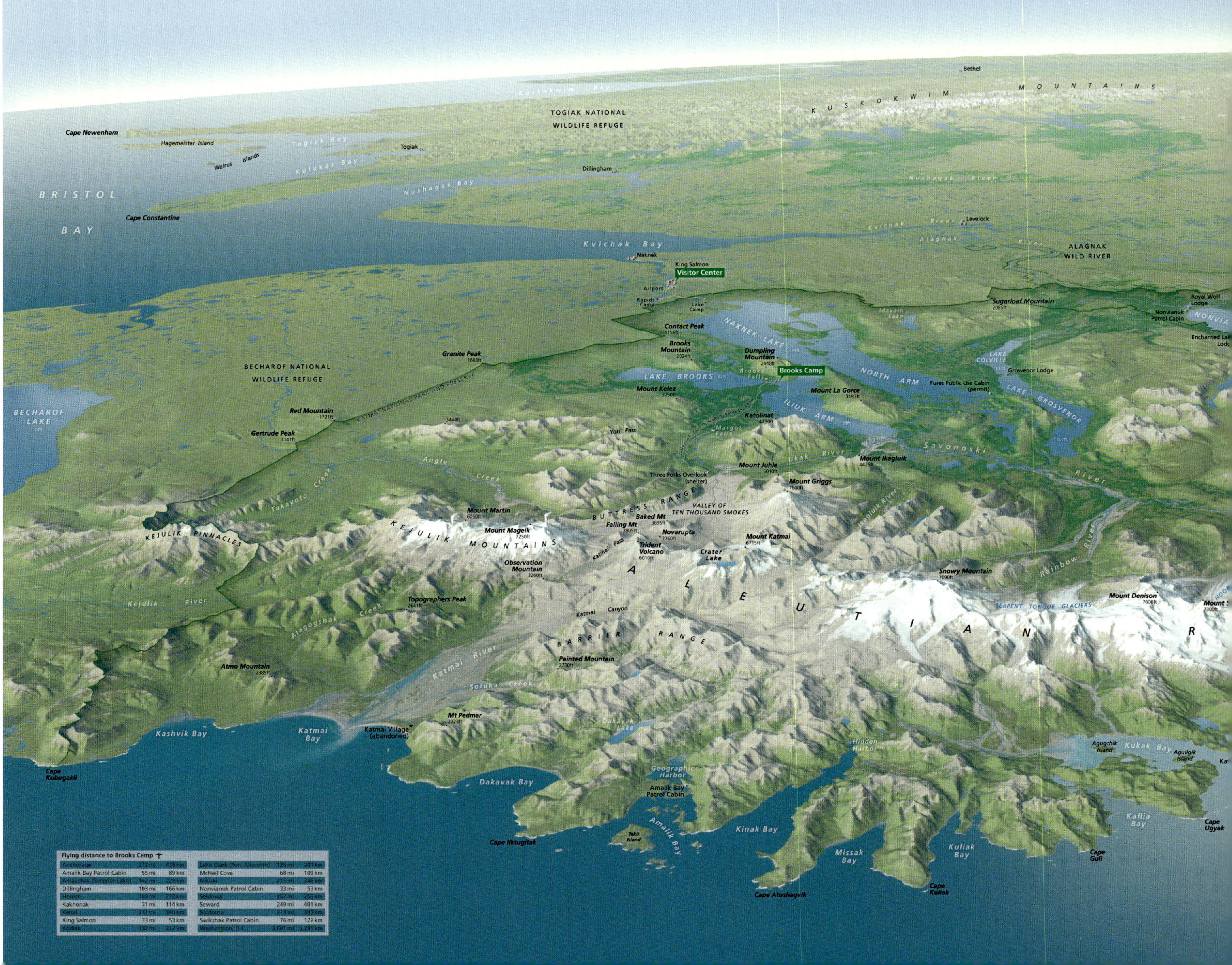

Flying distance to Brooks Camp

Anchorage	272 mi	438 km	Lake Clark (Port Alsworth)	125 mi	201 km
Amalik Bay Patrol Cabin	55 mi	89 km	McNeil Cove	68 mi	109 km
Aniakchak (Surprise Lake)	142 mi	229 km	Nikiski	215 mi	346 km
Dillingham	103 mi	166 km	Nonvianuk Patrol Cabin	33 mi	53 km
Homer	169 mi	272 km	Seldovia	157 mi	253 km
Kakhonak	71 mi	114 km	Seward	249 mi	401 km
Kenai	211 mi	340 km	Soldotna	213 mi	343 km
King Salmon	33 mi	53 km	Swikshak Patrol Cabin	76 mi	122 km
Kodiak	132 mi	212 km	Washington, D.C.	3,601 mi	5,795 km

ark and Preserve

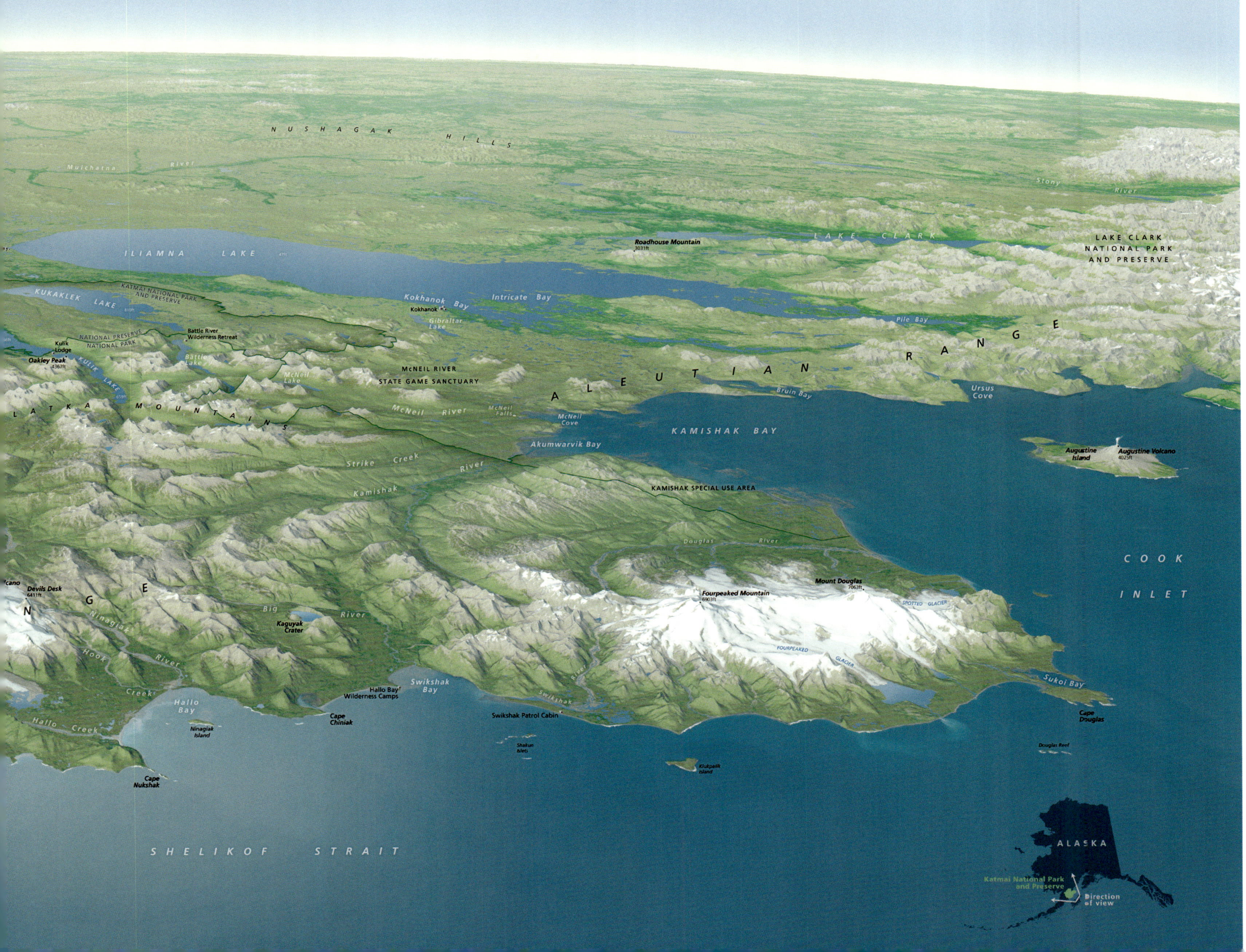

《 Katmai National Park and Preserve Panorama

Joe Milbrath and Jim Eynard

Wrangell-St. Elias National Park and Preserve 》

Tom Patterson

Note: This essay was written jointly by all cartographers (Milbrath, Eynard, and Patterson).

The National Parks of Alaska are home to some of the most spectacular terrain in North America. Yet because of their remoteness, vast size, and limited tourist infrastructure, few people visit the parks, and when they do, they explore only at a superficial level. These maps of Katmai National Park and Preserve and Wrangell-St. Elias National Park and Preserve offer 3D panoramic views to reveal the immense glaciers, volcanic terrain, and boreal vegetation of these natural treasures.

The panorama of Katmai National Park and Preserve highlights areas far from the popular Brooks Camp, where visitors go to watch bears. The rugged Aleutian Range arcs across the map, with a series of lakes and bays jutting perpendicularly into the mountains. The result is a study in contrasts: blue water intermingling with dense green forests, giving way to tundra, bare rock, and glaciated peaks as elevations increase. We used a Landsat image mosaic to show river-borne sediments reaching into Shelikof Strait. Hand-drawn elements, including waves breaking along the coastline, steam plumes venting from active volcanic peaks, and sunshine glinting off lakes and bays, enhance the map with movement and life.

Wrangell-St. Elias is a place of superlatives. It is the largest U.S. National Park and about the size of Switzerland. The mountains hold more glacial ice than anywhere else outside of polar regions. And 18,008-foot Mount St. Elias is the highest coastal mountain on Earth. Depicting all of this on the park brochure required four maps, including a panorama of the Wrangell Mountains. The panorama offers visitors a glimpse of the mountains they cannot see driving the 60-mile dirt road to the town of McCarthy, because the view is obscured by trees and hills. Unlike the Katmai panorama, the Wrangell panorama does not include a sky. Instead, to make room for supporting text in the brochure layout, it fades softly into the conventional park map above it.

We're proud that our maps of Katmai and Wrangell-St. Elias can show people magnificent wild places they otherwise would never have a chance to see.

Learn more at nps.gov/carto

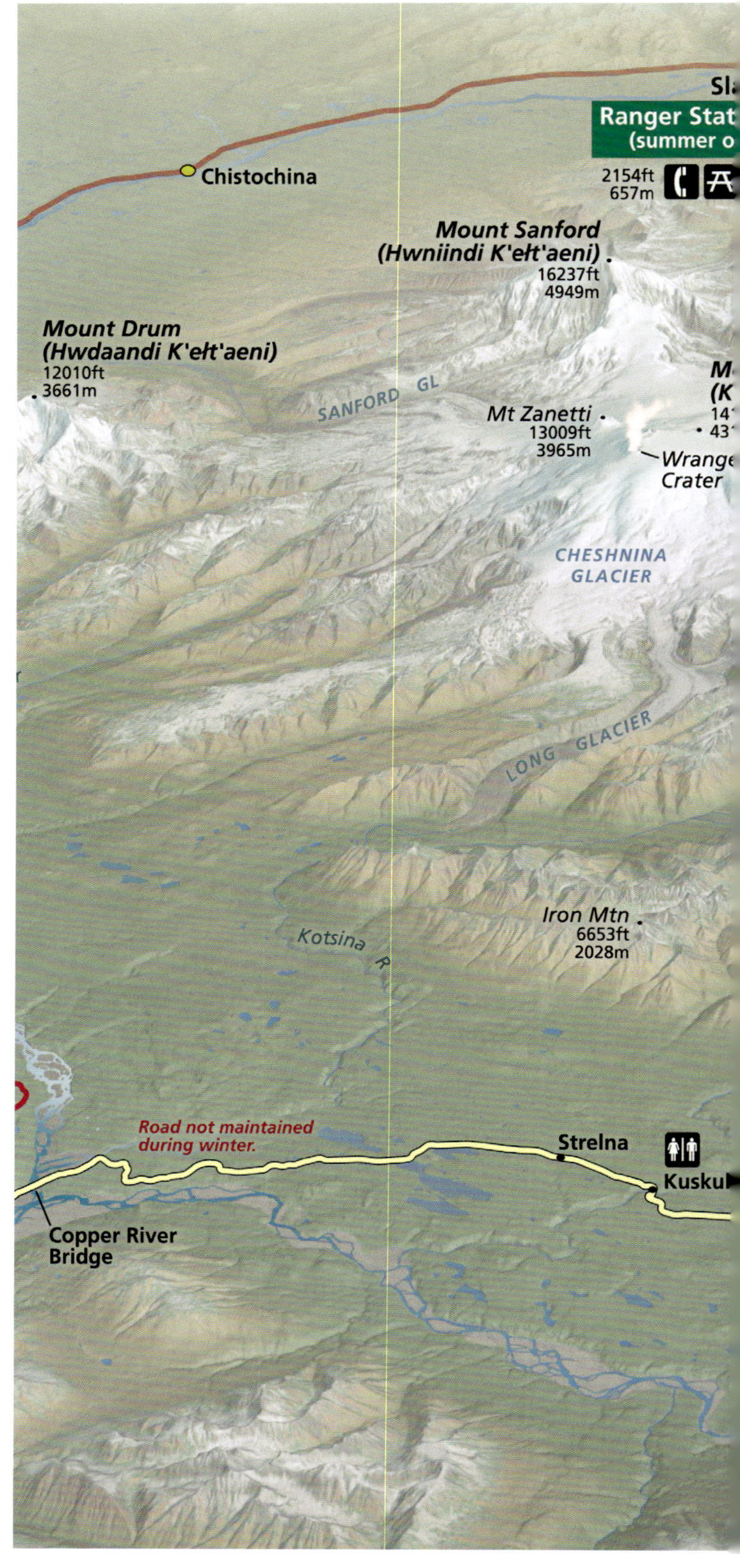

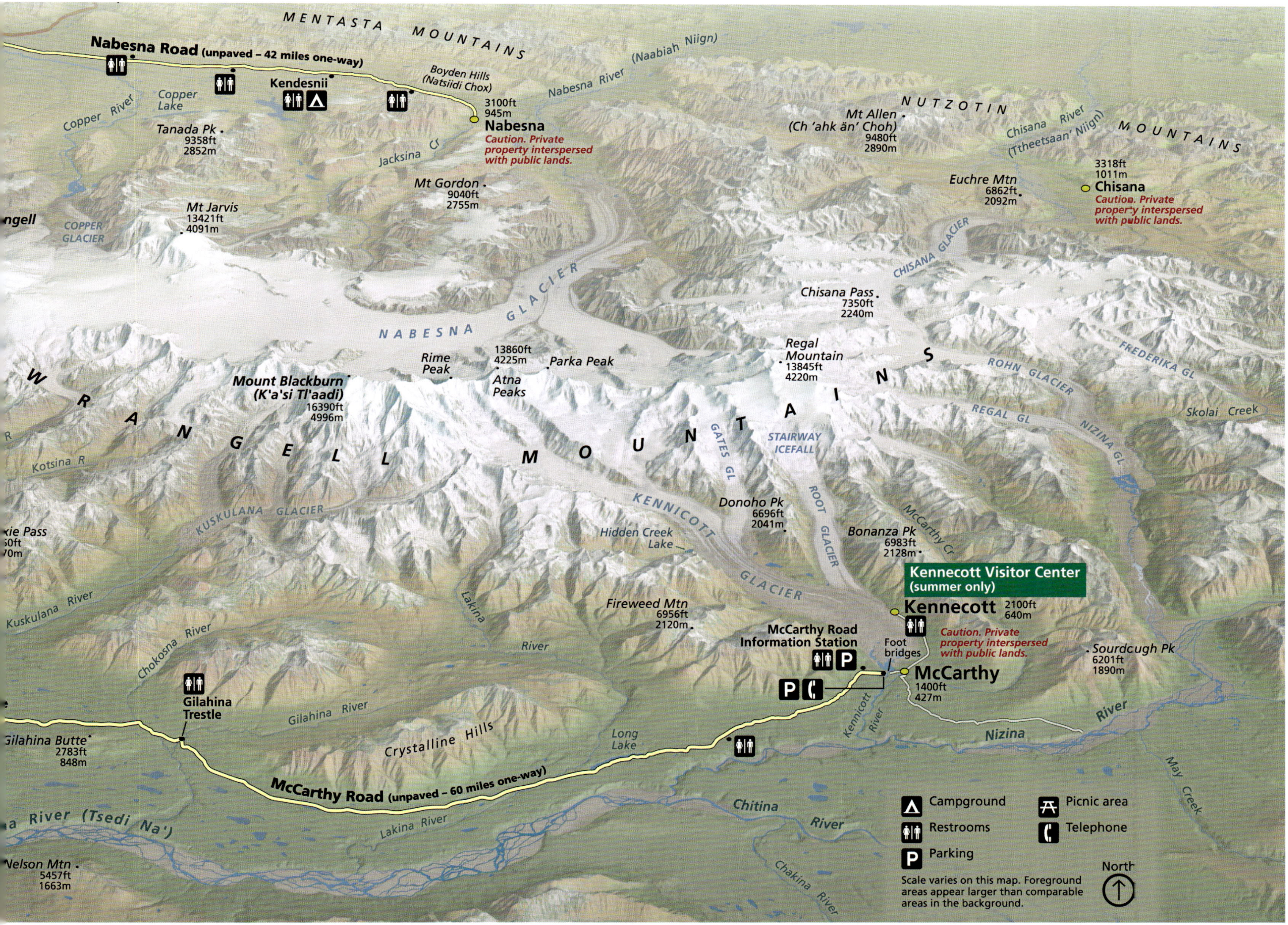

MENTASTA MOUNTAINS
Nabesna Road (unpaved – 42 miles one-way)
Kendesnii
Copper Lake
Copper River
Tanada Pk
9358ft
2852m
Boyden Hills
(Natsiidi Chox)
3100ft
945m
Nabesna
Caution. Private property interspersed with public lands.
Nabesna River
(Naabiah Niign)
Jacksina Cr
Mt Gordon
9040ft
2755m
Mt Jarvis
13421ft
4091m
COPPER GLACIER
NUTZOTIN MOUNTAINS
Mt Allen
(Ch 'ahk än' Choh)
9480ft
2890m
Chisana River
(Ttheetsaan' Niign)
3318ft
1011m
Chisana
Caution. Private property interspersed with public lands.
Euchre Mtn
6862ft
2092m
CHISANA GLACIER
NABESNA GLACIER
Chisana Pass
7350ft
2240m
Regal Mountain
13845ft
4220m
FREDERIKA GL
ROHN GLACIER
REGAL GL
Skolai Creek
NIZINA GL
Rime Peak
13860ft
4225m
Parka Peak
Atna Peaks
Mount Blackburn
(K'a'si Tl'aadi)
16390ft
4996m
WRANGELL MOUNTAINS
GATES GL
STAIRWAY ICEFALL
Kotsina R
KUSKULANA GLACIER
KENNICOTT GLACIER
ROOT GLACIER
Donoho Pk
6696ft
2041m
Hidden Creek Lake
Bonanza Pk
6983ft
2128m
McCarthy Cr
Kennecott Visitor Center
(summer only)
Kennecott
2100ft
640m
Caution. Private property interspersed with public lands.
Fireweed Mtn
6956ft
2120m
Lakina River
Kuskulana River
Chokosna River
McCarthy Road Information Station
Foot bridges
McCarthy
1400ft
427m
Sourdough Pk
6201ft
1890m
Kennicott River
Nizina River
Gilahina Trestle
Gilahina River
Gilahina Butte
2783ft
848m
Crystalline Hills
Long Lake
McCarthy Road (unpaved – 60 miles one-way)
Chitina River
(Tsedi Na')
Lakina River
May Creek
Chakina River
Nelson Mtn
5457ft
1663m
Campground
Picnic area
Restrooms
Telephone
Parking
North
Scale varies on this map. Foreground areas appear larger than comparable areas in the background.

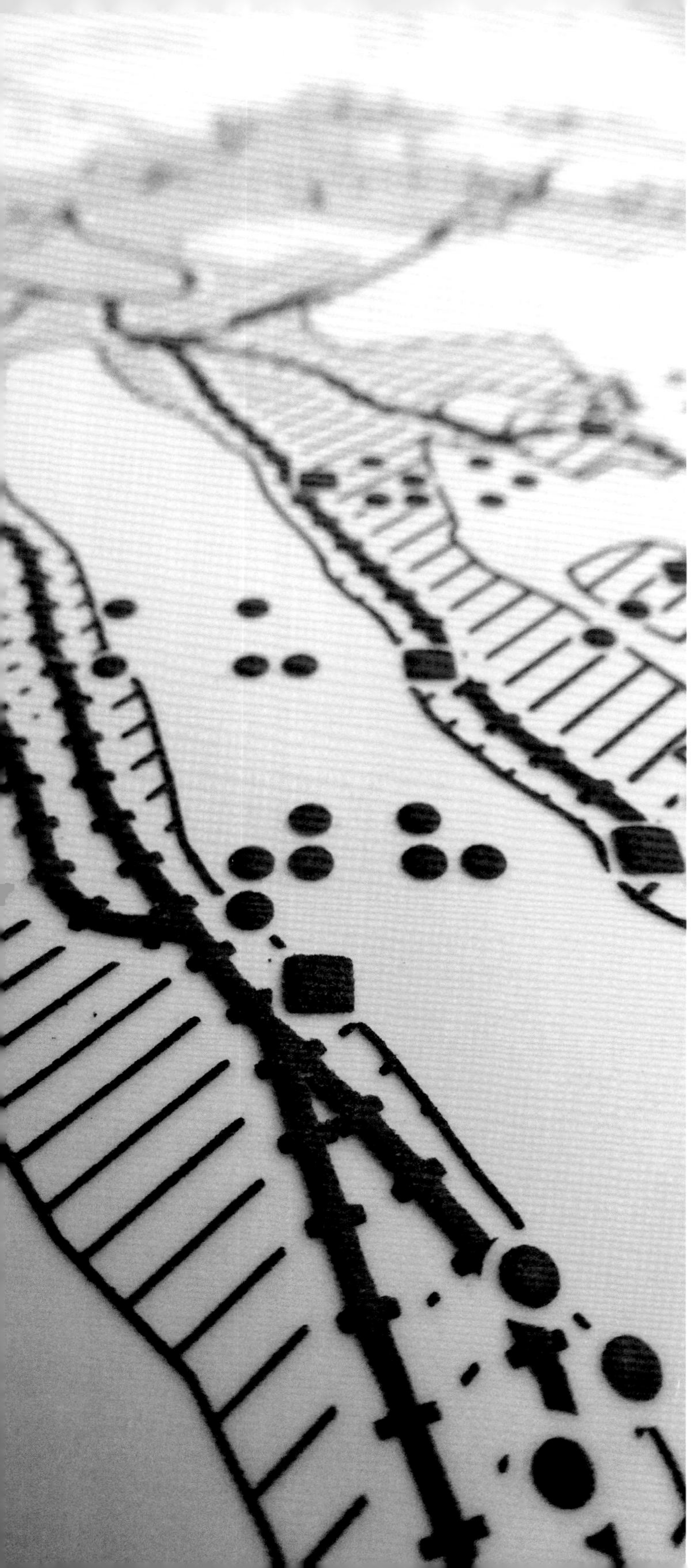

Tactile Map of Switzerland »

Anna Vetter

People with visual impairments have few opportunities to access geographic data through maps in order to better understand spatial reality. I first realized this problem when I was asked to create a tactile atlas of Switzerland. I had never before considered how maps could be designed to make them "readable" with one's fingertips.

Although the symbols are simple, the design process for this map was challenging. I always kept in mind that the surface would be touched rather than seen, so that typical cartographic rules would not apply. Symbols had to be bigger, more spread apart, and masked out heavily in relation to other symbols or labels. The labels in braille had to be in an unusually large size to be readable, and the limitations of space on the map required a two-letter abbreviation system with an accompanying abbreviation dictionary so that the audience could understand the map.

While creating the map, I engaged in a constant balancing of priorities, keeping the density of information as high as possible but as low as necessary to be easily readable. I made the unusual decision to depict the data at a larger scale (1:100,000) than the scale at which it was captured (1:500,000). I also split the themes into two different maps. One map includes settlements, administrative units, and the nation's rail network, and the other shows settlements, mountains, and waterways. To improve the audience's understanding of the overall geographic distribution, I also created some overview maps showing the themes.

The unique printing technique for these maps—a crucial component of the process—fascinated me. They are printed on special paper and then pushed through an oven. The paper swells where ink is applied as it reacts to the heat of the oven, creating a map on which all the features can be touched.

Making geography accessible to visually impaired people and sharing a sense of place with them in a new way has been deeply fulfilling. I hope this project and my methodology pave the way for the development of more tactile maps representing places and themes around the globe.

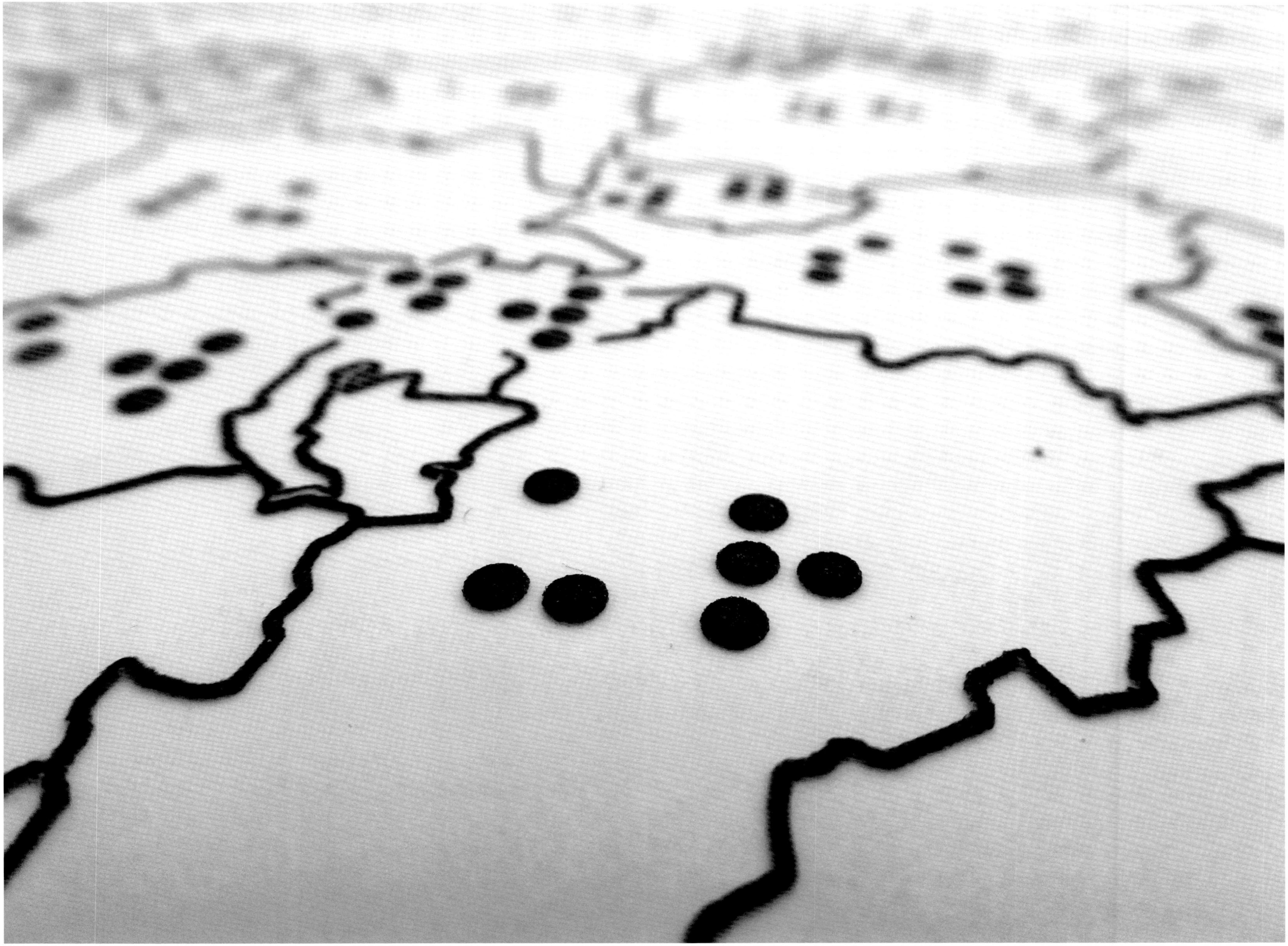

CANADIAN ROCKY MOUNTAIN PARKS MAP SERIES

BANFF, YOHO AND KOOTENAY NATIONAL PARKS

Go back 200 million years and the rocks that make up the Canadian Rocky Mountains are under water, part of a vast ocean floor. The jagged Rockies we see today are a result of intense geologic forces lifting the ocean beds that were later sculpted by icefields, glaciers and water. The area has experienced several ice ages over a period of two million years and has provided vast quantities of glacial water that has continuously nurtured life in the Canadian Rockies. However, in these times of unprecedented climate change, the area's glaciers and icefields are now in retreat.

The Athabasca Glacier, the largest glacier of the Columbia Icefield, has receded more than 1.5 kilometres and lost half of its volume in the last 125 years – in spite of receiving over seven metres of snow a year. The Peyto Glacier, one of the world's most-studied glaciers, is melting so quickly that scientists suggest it will lose 80–90% of its present day volume by 2100. The glaciers of the Rockies are not alone in their retreat; it is a global phenomenon. In fact, since 1960, glaciers worldwide have lost an estimated 8,000 cubic kilometres of ice (NASA, 2017).

In 1984, Banff, Jasper, Yoho and Kootenay National Parks, as well as the Mount Assiniboine, Mount Robson and Hamber provincial parks, are designated as World Heritage Sites by the United Nations Educational, Scientific and Cultural Organization (UNESCO). These parks made it on to the prestigious list because within the borders of the Canadian Rocky Mountain Parks are strikingly beautiful natural landscapes and the unique Burgess Shale, one of the world's most significant fossil beds, that offers an exquisitely preserved and fascinating window into pre-historic life on Earth.

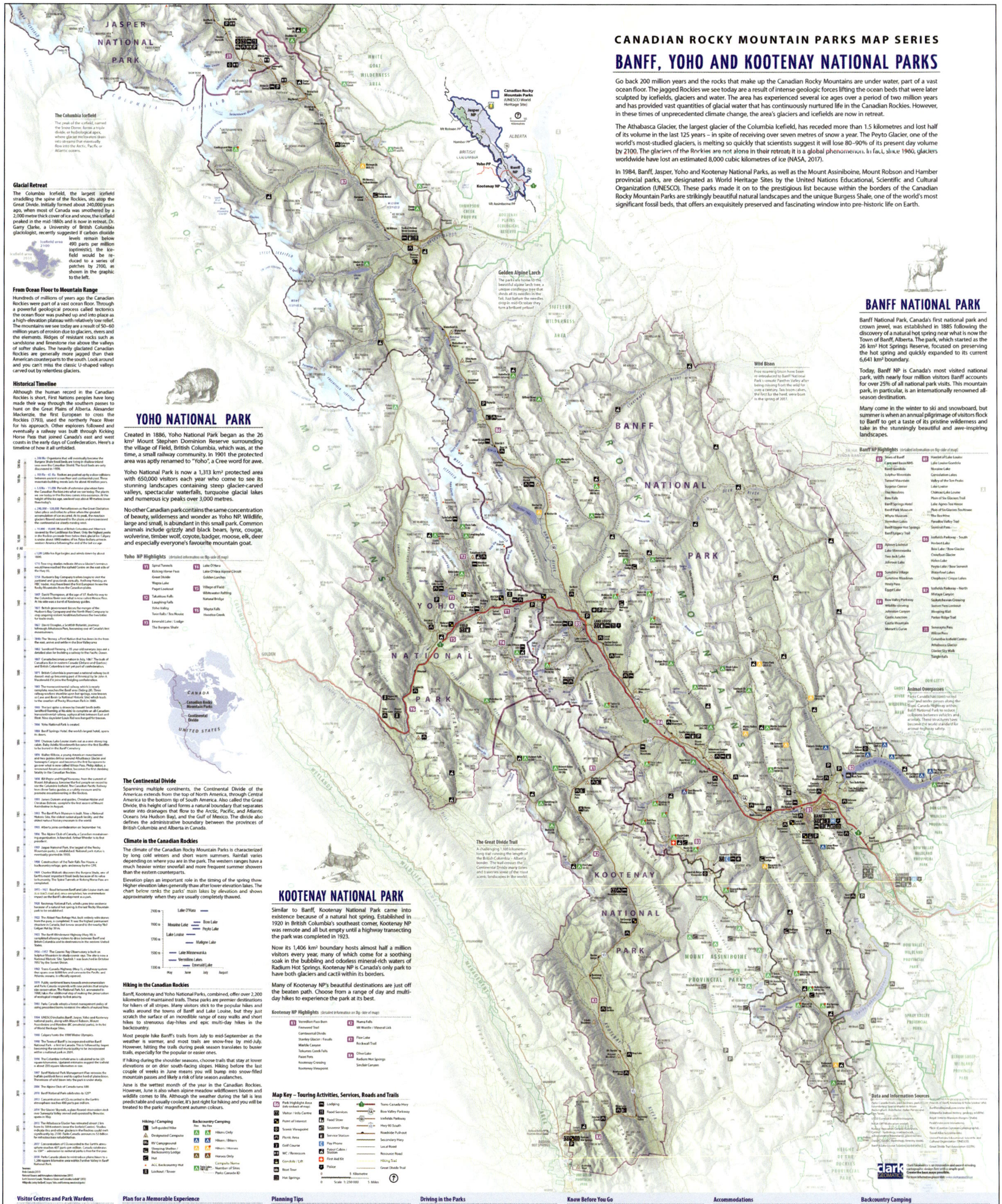

Glacial Retreat

The Columbia Icefield, the largest icefield straddling the spine of the Rockies, sits atop the Great Divide. Initially formed about 240,000 years ago, when most of Canada was smothered by a 2,000 metre thick cover of ice and snow, the icefield peaked in the mid-1880s and is now in retreat. Dr. Garry Clarke, a University of British Columbia glaciologist, recently suggested if carbon dioxide levels remain below 490 parts per million (optimistic), the icefield would be reduced to a series of patches by 2100, as shown in the graphic to the left.

From Ocean Floor to Mountain Range

Hundreds of millions of years ago the Canadian Rockies were part of a vast ocean floor. Through a powerful geological process called tectonics the ocean floor was pushed up and into place as a high-elevation plateau with relatively low relief. The mountains we see today are a result of 50–60 million years of erosion due to glaciers, rivers and the elements. Ridges of resistant rocks such as sandstone and limestone rise above the valleys of softer shales. The heavily glaciated Canadian Rockies are generally more jagged than their American counterparts to the south. Look around and you can't miss the classic U-shaped valleys carved out by relentless glaciers.

Historical Timeline

Although the human record in the Canadian Rockies is short, First Nations peoples have long made their way through the southern passes to hunt on the Great Plains of Alberta. Alexander Mackenzie, the first European to cross the Rockies (1793), used the northerly Peace River for his approach. Other explorers followed and eventually a railway was built through Kicking Horse Pass that joined Canada's east and west coasts in the early days of Confederation. Here's a timeline of how it all unfolded.

YOHO NATIONAL PARK

Created in 1886, Yoho National Park began as the 26 km² Mount Stephen Dominion Reserve surrounding the village of Field, British Columbia, which was, at the time, a small railway community. In 1901 the protected area was aptly renamed to "Yoho", a Cree word for awe.

Yoho National Park is now a 1,313 km² protected area with 650,000 visitors each year who come to see its stunning landscapes containing steep glacier-carved valleys, spectacular waterfalls, turquoise glacial lakes and numerous icy peaks over 3,000 metres.

No other Canadian park contains the same concentration of beauty, wilderness and wonder as Yoho NP. Wildlife, large and small, is abundant in this small park. Common animals include grizzly and black bears, lynx, cougar, wolverine, timber wolf, coyote, badger, moose, elk, deer and especially everyone's favourite mountain goat.

Yoho NP Highlights (detailed information on flip-side of map)

- Y1 Spiral Tunnels, Kicking Horse Pass, Great Divide, Wapta Lake, Paget Lookout
- Y2 Takakkaw Falls, Laughing Falls, Yoho Valley, Twin Falls / Tea House
- Y3 Emerald Lake / Lodge, The Burgess Shale
- Y4 Lake O'Hara, Lake O'Hara Alpine Circuit, Golden Larches
- Y5 Village of Field, Whitewater Rafting, Natural Bridge
- Y6 Wapta Falls, Hoodoo Creek

The Continental Divide

Spanning multiple continents, the Continental Divide of the Americas extends from the top of North America, through Central America to the bottom tip of South America. Also called the Great Divide, this height of land forms a natural boundary that separates water into drainages that flow to the Arctic, Pacific, and Atlantic Oceans (via Hudson Bay), and the Gulf of Mexico. The divide also defines the administrative boundary between the provinces of British Columbia and Alberta in Canada.

Climate in the Canadian Rockies

The climate of the Canadian Rocky Mountain Parks is characterized by long cold winters and short warm summers. Rainfall varies depending on where you are in the park. The western ranges have a much heavier winter snowfall and more frequent summer showers than the eastern counterparts.

Elevation plays an important role in the timing of the spring thaw. Higher elevation lakes generally thaw after lower elevation lakes. The chart below ranks the parks' main lakes by elevation and shows approximately when they are usually completely thawed.

Hiking in the Canadian Rockies

Banff, Kootenay and Yoho National Parks, combined, offer over 2,200 kilometres of maintained trails. These parks are premier destinations for hikers of all stripes. Many visitors stick to the popular hikes and walks around the towns of Banff and Lake Louise, but they just scratch the surface of an incredible range of easy walks and short hikes to strenuous day-hikes and epic multi-day hikes in the backcountry.

Most people hike Banff's trails from July to mid-September as the weather is warmer, and most trails are snow-free by mid-July. However, hitting the trails during peak season translates to busier trails, especially for the popular or easier ones.

If hiking during the shoulder seasons, choose trails that stay at lower elevations or on drier south-facing slopes. Hiking before the last couple of weeks in June means you will bump into snow-filled mountain passes and likely a risk of late season avalanches.

June is the wettest month of the year in the Canadian Rockies. However, June is also when alpine meadow wildflowers bloom and wildlife comes to life. Although the weather during the fall is less predictable and usually cooler, it's just right for hiking and you will be treated to the parks' magnificent autumn colours.

KOOTENAY NATIONAL PARK

Similar to Banff, Kootenay National Park came into existence because of a natural hot spring. Established in 1920 in British Columbia's southeast corner, Kootenay NP was remote and all but empty until a highway transecting the park was completed in 1923.

Now its 1,406 km² boundary hosts almost half a million visitors every year, many of which come for a soothing soak in the bubbling and odorless mineral-rich waters of Radium Hot Springs. Kootenay NP is Canada's only park to have both glaciers and cactii within its borders.

Many of Kootenay NP's beautiful destinations are just off the beaten path. Choose from a range of day and multi-day hikes to experience the park at its best.

Kootenay NP Highlights (detailed information on flip-side of map)

- K1 Vermilion Pass Burn, Fireweed Trail, Continental Divide, Stanley Glacier / Fossils, Marble Canyon, Tokumm Creek Falls, Paint Pots, Kootenay Crossing, Kootenay Viewpoint
- K2 Numa Falls, Mt Wardle / Mineral Lick
- K3 Floe Lake, Rockwall Trail
- K4 Olive Lake, Radium Hot Springs, Sinclair Canyon

Map Key – Touring Activities, Services, Roads and Trails

Hiking / Camping	Backcountry Camping	Touring	Services	Roads and Trails
Self-guided Hike	Hikers Only	Park Highlight Area (info on back of map)	Lodging	Trans-Canada Hwy
Designated Campsite	Hikers / Bikers	Visitor / Info Centre	Food Services	Bow Valley Parkway
RV Campground	Hikers / Horses	Point of Interest	Food Store	Icefields Parkway
Sleeping Shelter / Backcountry Lodge	Horses Only	Scenic Viewpoint	Souvenir Shop	Hwy 93 South
Hut	Campsite Name, Number of Sites, Parks Canada ID	Picnic Area	Service Station	Roadside Pull-out
ACC Backcountry Hut		Golf Course	Pay Phone	Secondary Hwy
Lookout / Tower		WC / Restroom	Patrol Cabin / Station	Local Road
		Gondola / Lift	First Aid Kit	Resource Road
		Boat Tour	Police	Hiking Trail
		Hot Springs		Great Divide Trail

Backcountry Camping: Fire / No Fire

Scale 1: 250 000 — 0–5 Kilometre; 0–5 Miles

BANFF NATIONAL PARK

Banff National Park, Canada's first national park and crown jewel, was established in 1885 following the discovery of a natural hot spring near what is now the Town of Banff, Alberta. The park, which started as the 26 km² Hot Springs Reserve, focused on preserving the hot spring and quickly expanded to its current 6,641 km² boundary.

Today, Banff NP is Canada's most visited national park, with nearly four million visitors Banff accounts for over 25% of all national park visits. This mountain park, in particular, is an internationally renowned all-season destination.

Many come in the winter to ski and snowboard, but summer is when an annual pilgrimage of visitors flock to Banff to get a taste of its pristine wilderness and take in the stunningly beautiful and awe-inspiring landscapes.

Banff NP Highlights (detailed information on flip-side of map)

- B1 Town of Banff, Cave and Basin NHS, Banff Gondola, Sulphur Mountain, Tunnel Mountain, Surprise Corner, The Hoodoos, Bow Falls, Banff Springs Hotel, Banff Park Museum, Whyte Museum, Vermilion Lakes, Banff Upper Hot Springs, Banff Legacy Trail
- B5 Hamlet of Lake Louise, Lake Louise Gondola, Moraine Lake, Consolation Lakes, Valley of the Ten Peaks, Lake Louise, Château Lake Louise, Plain of Six Glaciers Trail, Lake Agnes Tea House, Plain of Six Glaciers Tea House, The Bee Hive, Paradise Valley Trail, Sentinel Pass
- B6 Icefields Parkway – South, Herbert Lake

Visitor Centres and Park Wardens

Five **official park visitor centres, located in Banff, Lake Louise, Radium,** and **Field** are where you'll find very knowledgeable Parks Canada staff. Ask them about park activities, accommodations, places to eat, bus schedules, special events, and more. Contact a **park warden** to report carnivore sightings, feeding and harrassment of wildlife, campsite noise, poaching, polluting and littering, and fishing violations.

1. **Banff Visitor Centre** — 224 Banff Avenue, Banff, AB
2. **Banff Visitor Information Kiosk** — Banff Heritage Railway Station, 327 Railway Avenue, Banff, AB
3. **Lake Louise Visitor Centre** — Village of Lake Louise, AB (located next to Samson Mall)
4. **Field Visitor Centre** — 5764 Trans-Canada Hwy, Field, BC
5. **Kootenay Visitor Centre** — 7556 Main St. East, Radium, BC
6. **Park Wardens** — Call toll-free, 24/7

Plan for a Memorable Experience

The Canadian Rocky Mountain Parks are easily reached from the western cities of **Calgary**, **Edmonton** and **Vancouver** by car, bus or train. While the parks are mostly vast areas of wilderness waiting to be explored, there is a network of highways and full-service towns that easily accommodate most visitor needs. With so much to do within these vast parks, and because you can't do it all, you need to spend your precious time wisely.

Every year millions of people travel through these parks to see the natural beauty of the landscape. Although the parks are visited year-round, the **best time is between June and September**, which means you'll experience crowds at popular destinations and scenic viewpoints during the peak months of July and August. Most sites and attractions start operating in early June through to mid-October.

While many visitors barely stop as they make their way up and down the busy Banff to Lake Louise corridor visiting one crowded site after another, it doesn't have to be that way. A well-planned four or five day visit in the parks gives you time to absorb the highlights and experience the vast and scenic wilderness within these parks (but longer is better).

Planning Tips

1. **Use this map.** Each park has unique attractions and points of interest. To help plan your trip this map divides each park into areas – indicated by one of these symbols Y6 – and lists highlights to check out in each area.
2. Arrive at popular destinations well before 9 AM or in the early evening to avoid most of the crowds during the peak season (perk: you'll have better lighting for your landscape and wildlife photos).
3. Give yourself enough time to travel to your destination – especially along the busy Icefield Parkway and Bow Valley Parkway routes where there are several roadside pull-outs with scenic views, roaming wildlife and crowds.
4. Plan for one major activity per day but remember to also put aside some time for randomly exploring or simply enjoying a riverside picnic. The staff at the Visitor Centres can point you in the right direction. Check for seasonal and safety closures on the Parks Canada website to avoid disappointment.
5. Get off the beaten track by taking one of many trails that start at a roadside viewpoint. Any longer and tougher hike will allow you to escape from the crowds and your efforts will be rewarded with fantastic views.
6. Consider visiting the parks in the calmer shoulder seasons (spring and fall), or even during the winter, for a truly unique experience.

Driving in the Parks

The Canadian Rocky Mountain Parks are perfect for **scenic driving tours**. A well-maintained highway and parkway system takes you deep into the glaciated landscape providing easy access to emerald-coloured lakes, forested valleys, wetlands, alpine meadows, glaciers and a chance to see some spectacular wildlife. There are several drives to choose from: the world famous **Icefields Parkway**; the wildlife-viewer's **Bow Valley Parkway**; the **Lake Minnewanka Loop**; and the **Vermilion Lakes Drive**.

1. Obey posted speed limits: 90 km/hr (56 mph) on major routes and 60 km/hr (37 mph) on secondary roads. Tickets are pricey, so pay attention.
2. Be prepared for a variety of driving conditions. Higher elevations can get snow in the summer months. Winter driving conditions can be treacherous – equip your car with food, extra warm clothes and blankets (or a sleeping bag), flashlight and a shovel.
3. Drive defensively and keep an eye out for wildlife, especially at dusk/dawn when lighting is poor and animal activity is at its highest.
4. When traveling on highways, including the Trans-Canada (Highway 1), your vehicle is required by law to have winter tires or mud and snow (M+S) tires from October 1st through to March 31st.

Know Before You Go

1. Entry to Canada's national parks requires a valid **Park Pass** (valid for all national parks, marine conservation areas and historic sites). Daily and Annual passes available for purchase at park gates and information centres.
2. Visitor centres are located in all four national parks. Make sure you stop by and talk to the knowledgeable staff - they actually have all the answers!
3. Emergency: **call 9-1-1 for police, fire and ambulance.**
4. Cell phone reception within the parks is limited and unreliable at best.
5. Pay phones are available at visitor centres and service centres.
6. There is no internet or WiFi in the parks, except for the usual suspects like coffee shops, and restaurants within the towns and service centres.
7. Pets are welcome but they must be on a leash at all times. Pets are not allowed in backcountry shelters.
8. Check the opening dates for campsites with Parks Canada.
9. Bear warnings, closures and other hiker restrictions commonly occur throughout the park. Obtain up-to-date information by contacting a Parks Canada Visitor Centre.
10. If you are planning to hike, carry bear spray and know how to use it.

Accommodations

A wide range of accommodation is available in Banff, Lake Louise, Jasper, Field, Radium Hot Springs, Saskatchewan Crossing and nearby Canmore. Whether you decide to stay at a luxury hotel, a motel, a B&B, a simple hostel or a backcountry lodge you must **book well in advance (at least several months) to reserve**, otherwise you will be disappointed.

Visit websites such as Expedia, TripAdvisor, and BanffHotels.org for comprehensive listings of accommodations in the town of Banff, Lake Louise, and throughout the Rocky Mountain Parks and surrounding areas.

Camping

Campsite reservations are a must (where available) if you plan on camping in the peak summer months. Campsites requiring reservations and can be made up to three months in advance. The Parks Canada reservation system for the upcoming season is usually available in early January.

1. **Reserve by phone:** Call **1-877-RESERVE** (1.877.737.3783). Available from 8 AM to 6 PM local park time.
2. **Reserve online with Parks Canada:** reservation.pc.gc.ca

Backcountry Camping

Reservations and a **Wilderness Pass** are required for backcountry camping. Purchase a Wilderness Pass from a **Parks Canada Visitor Centre** for overnight trips (required any time of year).

When purchasing a Wilderness Pass, indicate which trails and campsites you will use, along with specific dates. Daily or annual passes are available and are valid for one year from date of purchase.

Mountain Bikers / Cyclists

Cyclists and mountain bikers are susceptible to sudden, dangerous bear encounters – slow down, stay alert and make noise. Please dismount when approaching horses and hikers on trails.

Horseback Riding

Horses were used in the Canadian Rockies by First Nations for hunting and trading. Traveling by horse is still permitted in most areas within the parks although some trails are restricted to reduce the impact on the vegetation and infrastructure. Horse rentals and guided pack trips are available from many outfitters. It's a great experience.

I wanted to evoke a sense of awe for the beauty and vastness of the wild landscape.

« Banff, Yoho & Kootenay National Parks Recreation Map and Visitor Guide

Jeff Clark — Clark Geomatics

In designing a map for some of the world's most magnificent places—Banff, Yoho, and Kootenay National Parks—I wanted to evoke a sense of awe for the beauty and vastness of the wild landscape. I also knew it was vital to provide the most relevant and helpful information to ensure the best trip possible for visitors to these parks in the Canadian Rocky Mountains.

To advance geographic interest in and awareness of these grand wilderness parks, and after extensive research of existing maps of the Rockies, I decided to create a fresh new map that is beautiful, geographically expressive, and tells a story. I focused on visual clarity and precise communication of details to help readers engage with the map and to provide a big picture view of all three parks.

The natural color scheme, derived from recent satellite imagery, resembles what you would see if you were flying over the park in an airplane. The natural look of the landscape gives readers a more realistic view of the mountains, glaciers, valleys, and lakes. Text surrounding the map explains the parks' natural history, geology, landscapes, climate, wildlife, and attractions. Well-placed and easy-to-read symbols and labels for hiking, camping, and points of interest make it simple for visitors to find the information they need. The final product is a rich trip-planning tool that can enhance every park visitor's experience.

Learn more at clarkgeomatics.ca

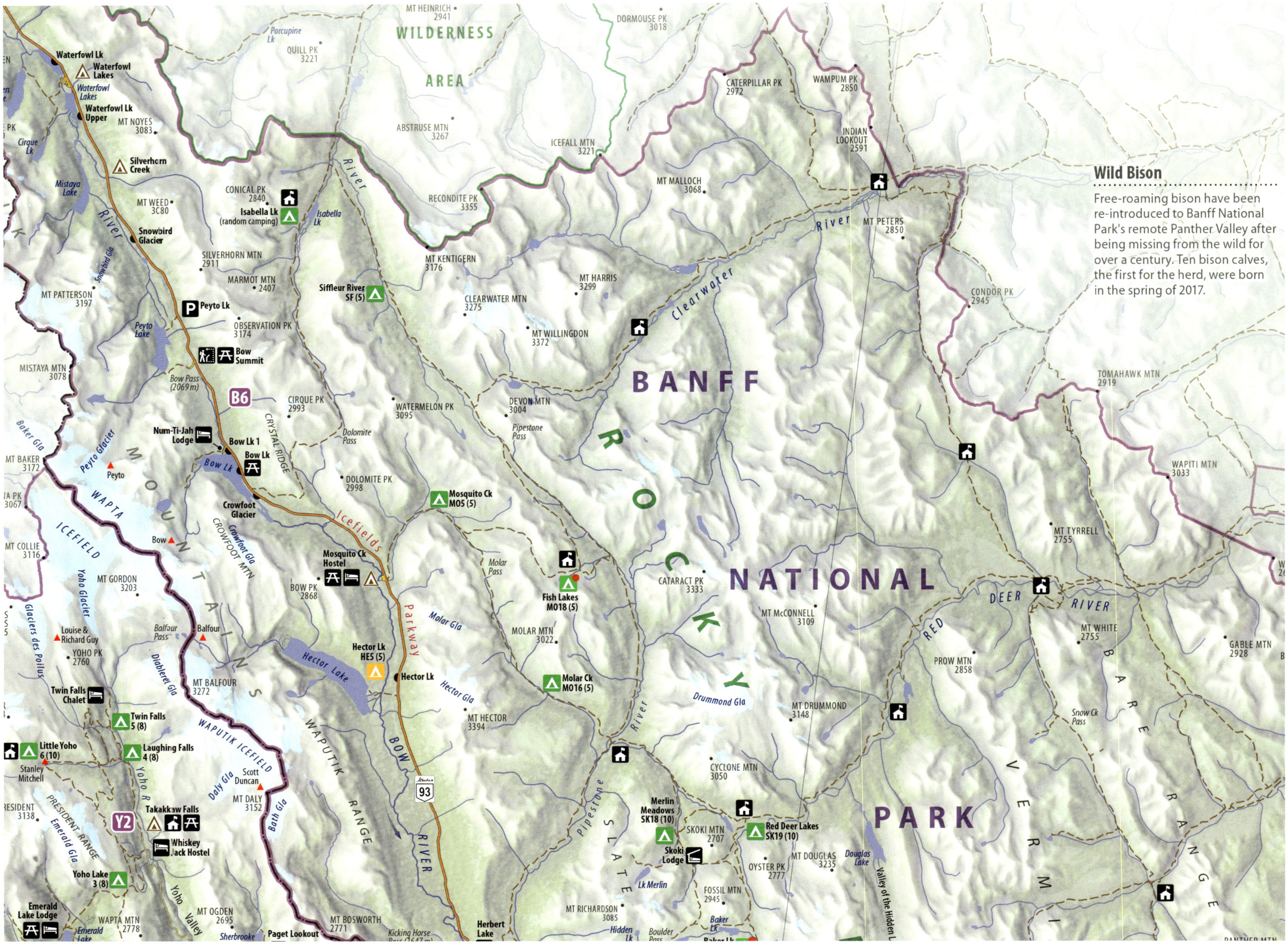

Wild Bison

Free-roaming bison have been re-introduced to Banff National Park's remote Panther Valley after being missing from the wild for over a century. Ten bison calves, the first for the herd, were born in the spring of 2017.

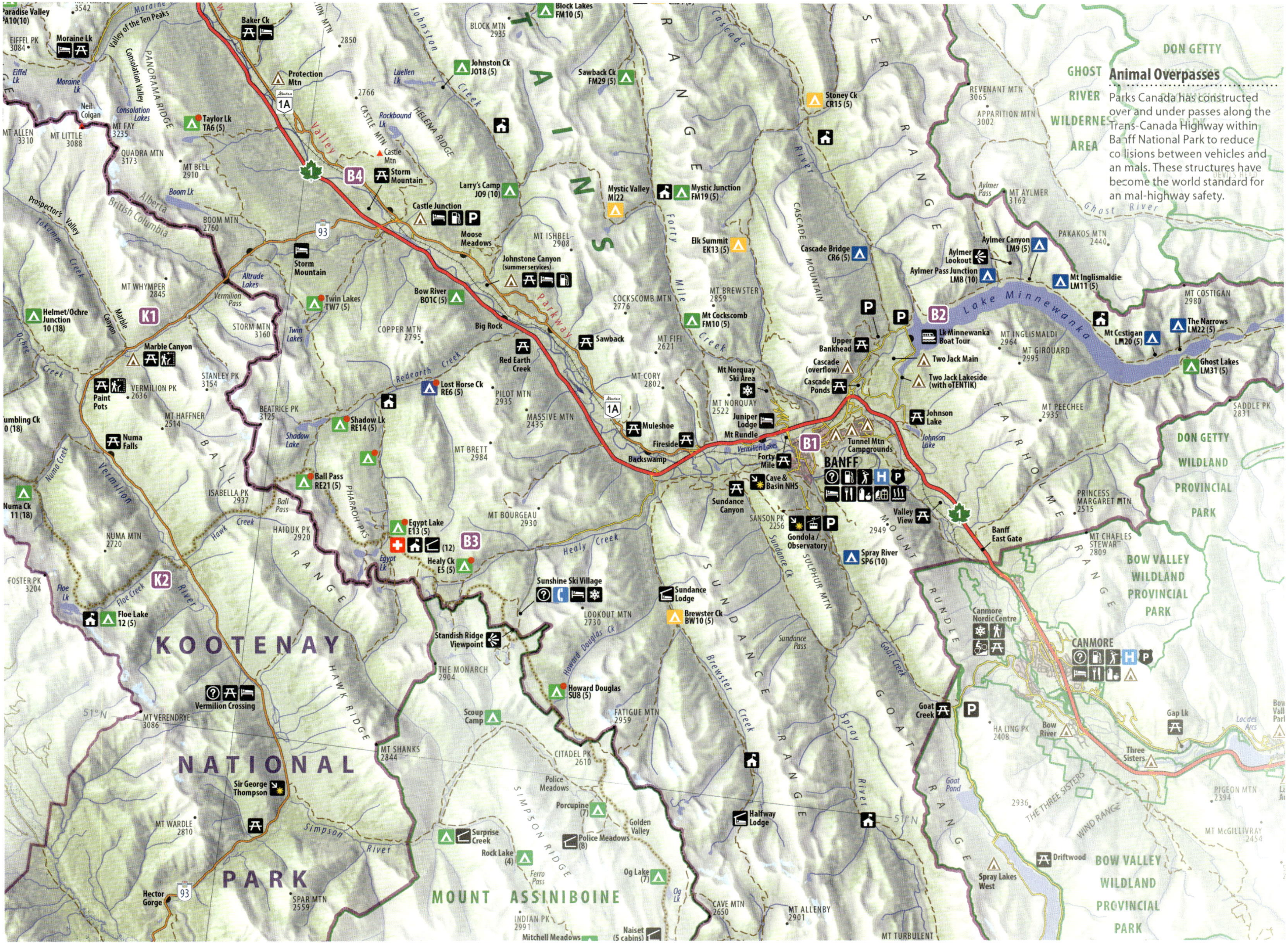

Animal Overpasses

Parks Canada has constructed over and under passes along the Trans-Canada Highway within Banff National Park to reduce collisions between vehicles and animals. These structures have become the world standard for animal-highway safety.

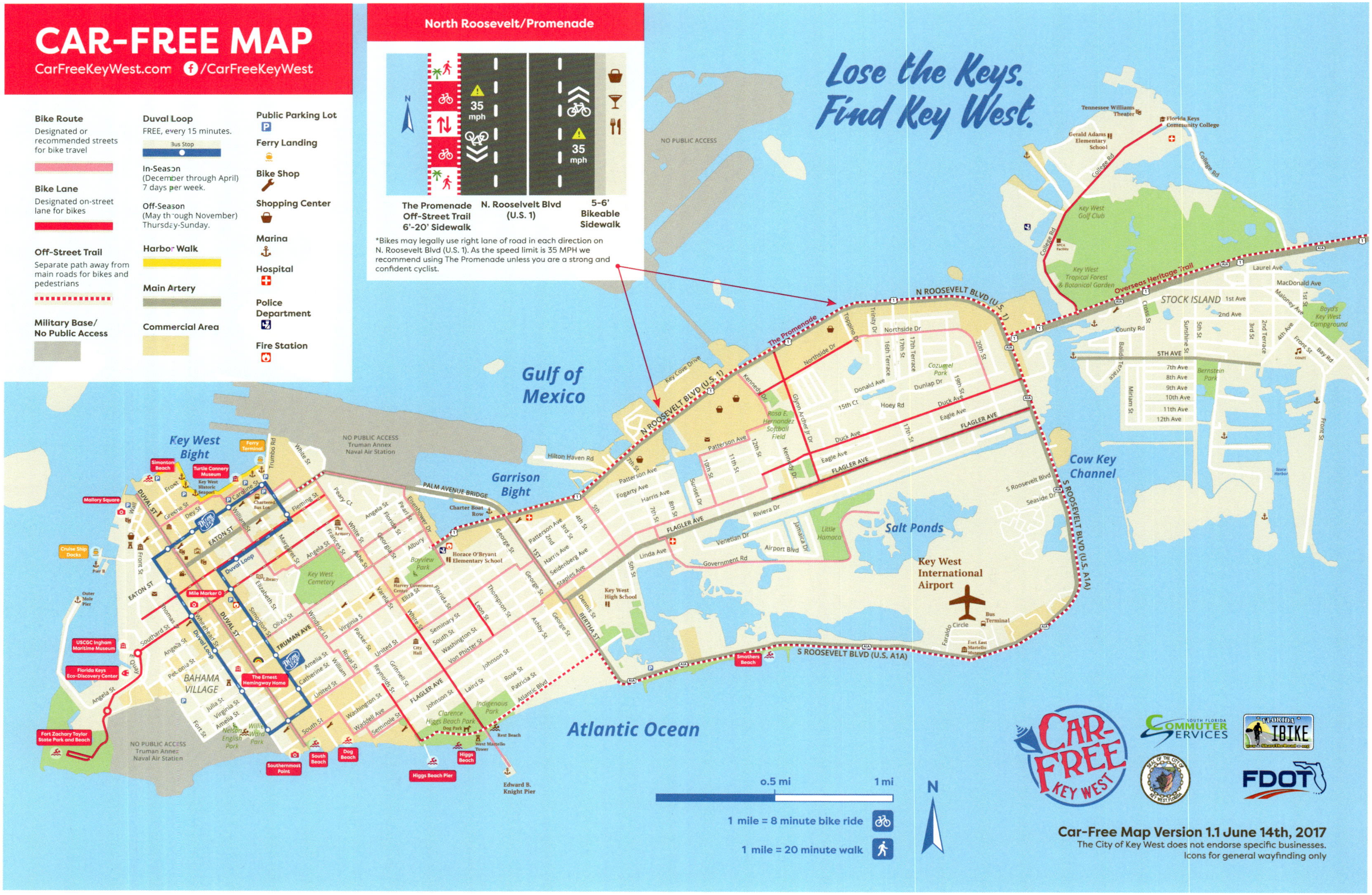
CAR-FREE MAP
CarFreeKeyWest.com /CarFreeKeyWest
Bike Route
Designated or recommended streets for bike travel
Bike Lane
Designated on-street lane for bikes
Off-Street Trail
Separate path away from main roads for bikes and pedestrians
Military Base/ No Public Access
Duval Loop
FREE, every 15 minutes.
Bus Stop
In-Season (December through April) 7 days per week.
Off-Season (May through November) Thursday-Sunday.
Harbor Walk
Main Artery
Commercial Area
Public Parking Lot
Ferry Landing
Bike Shop
Shopping Center
Marina
Hospital
Police Department
Fire Station
North Roosevelt/Promenade
35 mph
The Promenade Off-Street Trail 6'-20' Sidewalk
N. Roosevelt Blvd (U.S. 1)
5-6' Bikeable Sidewalk
*Bikes may legally use right lane of road in each direction on N. Roosevelt Blvd (U.S. 1). As the speed limit is 35 MPH we recommend using The Promenade unless you are a strong and confident cyclist.
Lose the Keys. Find Key West.
Gulf of Mexico
Key West Bight
Garrison Bight
Cow Key Channel
Salt Ponds
Atlantic Ocean
STOCK ISLAND
Key West International Airport
BAHAMA VILLAGE
N ROOSEVELT BLVD (U.S. 1)
S ROOSEVELT BLVD (U.S. A1A)
FLAGLER AVE
TRUMAN AVE
DUVAL ST
EATON ST
Overseas Heritage Trail
The Promenade
NO PUBLIC ACCESS
NO PUBLIC ACCESS Truman Annex Naval Air Station
Mallory Square
Fort Zachary Taylor State Park and Beach
Southernmost Point
Higgs Beach Pier
Smathers Beach
0.5 mi
1 mi
1 mile = 8 minute bike ride
1 mile = 20 minute walk
N
CAR-FREE KEY WEST
South Florida Commuter Services
FDOT
Car-Free Map Version 1.1 June 14th, 2017
The City of Key West does not endorse specific businesses. Icons for general wayfinding only

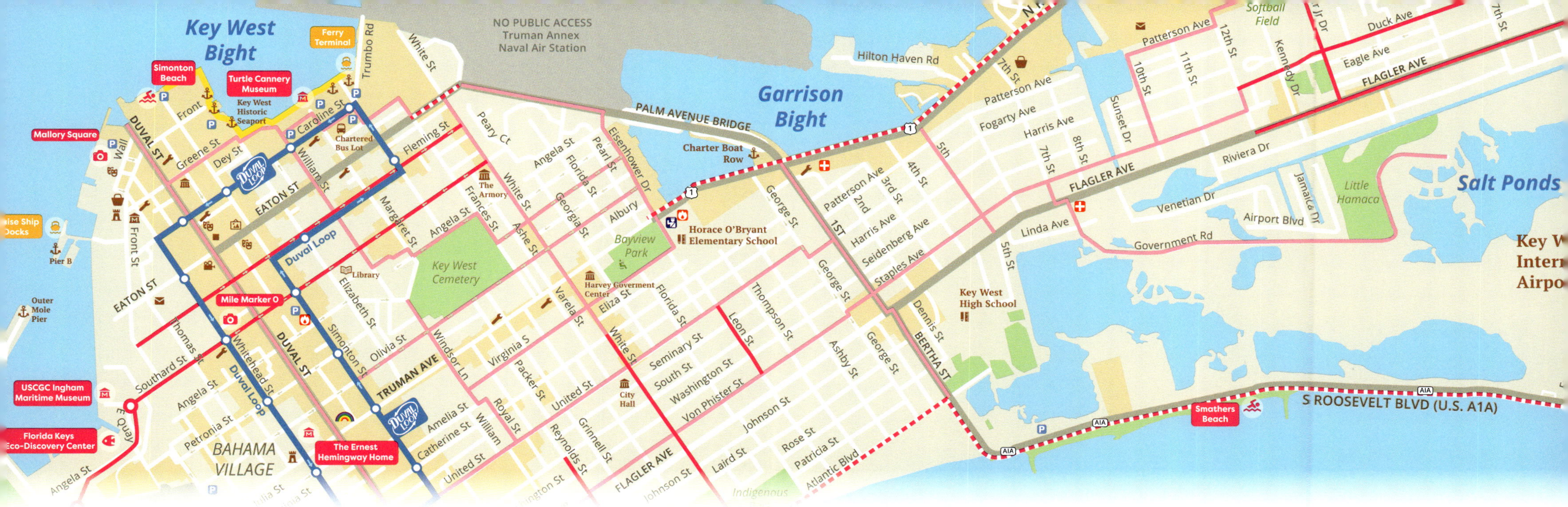

« Car-Free Key West

Kate Chanba for The City of Key West

We embraced a collaborative, hands-on approach in the map design process for the Car-Free Key West campaign. Our goal was to make this transportation map the default choice for the island—the best map to hand out at hotels, airports, and other tourist touchpoints. The mapping effort was coupled with a large-scale branding and outreach campaign that encouraged tourists to leave their cars behind when visiting, so branding was a major consideration in our design. The fun typefaces, bright colors, and modern icons influenced the creation of the logo and supporting brand materials for the broader campaign.

The process kicked off with intensive user research. My team conducted community interviews, biked the roads, and developed quick prototypes to get an on-the-ground understanding of a user's many options for traveling in Key West. We included hotels, restaurants, and bike rental locations in our research. And we facilitated sessions with project partners to define our communication strategies and potential challenges. One unique challenge was prioritizing safety while acknowledging the festive, carefree culture of the island.

The visual style of the map is designed to serve tourists. Places of interest are at the top of the hierarchy, facilitating point-to-point routing. A schematic map (not shown in this version) proved to be the most effective way to convey information about bus services without cluttering the main map with multi-colored transit routes, and the ubiquitous flamingo pink bike routes tell our audience that biking can get you just about anywhere you need to go. We included biking and walking times on our one-mile scale to emphasize how easy it is to "Lose the keys. Find Key West!"